CAMBRIDGE LIBRARY COLLECTION

Books of enduring scholarly value

Mathematical Sciences

From its pre-historic roots in simple counting to the algorithms powering
modern desktop computers, from the genius of Archimedes to the genius of
Einstein, advances in mathematical understanding and numerical techniques
have been directly responsible for creating the modern world as we know
it. This series will provide a library of the most influential publications and
writers on mathematics in its broadest sense. As such, it will show not only
the deep roots from which modern science and technology have grown, but
also the astonishing breadth of application of mathematical techniques in the
humanities and social sciences, and in everyday life.

The Mathematical Analysis of Logic

Self-taught mathematician George Boole (1815-64) published a pamphlet in
1847 – The Mathematical Analysis of Logic – that launched him into history
as one of the nineteenth century's most original thinkers. In the introduction,
Boole closely adheres to two themes: the fundamental unity of all science and
the close relationship between logic and mathematics. In the first chapter, he
examines first principles of formal logic, and then moves on to Aristotelian
syllogism, hypotheticals, and the properties of elective functions. Boole
uses this pamphlet to answer a well-known logician of the day, Sir William
Hamilton, who believed that only philosophers could study 'the science
of real existence', while all mathematicians could do was measure things.
In essence, The Mathematical Analysis of Logic humbly chides Hamilton
and asks him to rethink his bias. Boole is compelling reading for anyone
interested in intellectual history and the science of the mind.

Cambridge University Press has long been a pioneer in the reissuing of out-of-print titles from its own backlist, producing digital reprints of books that are still sought after by scholars and students but could not be reprinted economically using traditional technology. The Cambridge Library Collection extends this activity to a wider range of books which are still of importance to researchers and professionals, either for the source material they contain, or as landmarks in the history of their academic discipline.

Drawing from the world-renowned collections in the Cambridge University Library, and guided by the advice of experts in each subject area, Cambridge University Press is using state-of-the-art scanning machines in its own Printing House to capture the content of each book selected for inclusion. The files are processed to give a consistently clear, crisp image, and the books finished to the high quality standard for which the Press is recognised around the world. The latest print-on-demand technology ensures that the books will remain available indefinitely, and that orders for single or multiple copies can quickly be supplied.

The Cambridge Library Collection will bring back to life books of enduring scholarly value (including out-of-copyright works originally issued by other publishers) across a wide range of disciplines in the humanities and social sciences and in science and technology.

The Mathematical Analysis of Logic

Being an Essay Towards a Calculus of Deductive Reasoning

GEORGE BOOLE

CAMBRIDGE
UNIVERSITY PRESS

CAMBRIDGE UNIVERSITY PRESS

Cambridge, New York, Melbourne, Madrid, Cape Town, Singapore,
São Paolo, Delhi, Dubai, Tokyo

Published in the United States of America by Cambridge University Press, New York

www.cambridge.org
Information on this title: www.cambridge.org/9781108001014

© in this compilation Cambridge University Press 2009

This edition first published 1847
This digitally printed version 2009

ISBN 978-1-108-00101-4 Paperback

THE MATHEMATICAL ANALYSIS

OF LOGIC,

BEING AN ESSAY TOWARDS A CALCULUS
OF DEDUCTIVE REASONING.

BY GEORGE BOOLE.

Ἐπικοινωνοῦσι δὲ πᾶσαι αἱ ἐπιστῆμαι ἀλλήλαις κατὰ τὰ κοινά. Κοινὰ δὲ λέγω, οἷς χρῶνται ὡς ἐκ τούτων ἀποδεικνύντες· ἀλλ' οὐ περὶ ὧν δεικνύουσιν, οὐδε ὃ δεικνύουσι.

ARISTOTLE, *Anal. Post.*, lib. I. cap. XI.

CAMBRIDGE:
MACMILLAN, BARCLAY, & MACMILLAN;
LONDON: GEORGE BELL.

1847

PREFACE.

In presenting this Work to public notice, I deem it not irrelevant to observe, that speculations similar to those which it records have, at different periods, occupied my thoughts. In the spring of the present year my attention was directed to the question then moved between Sir W. Hamilton and Professor De Morgan; and I was induced by the interest which it inspired, to resume the almost-forgotten thread of former inquiries. It appeared to me that, although Logic might be viewed with reference to the idea of quantity,* it had also another and a deeper system of relations. If it was lawful to regard it from *without,* as connecting itself through the medium of Number with the intuitions of Space and Time, it was lawful also to regard it from *within,* as based upon facts of another order which have their abode in the constitution of the Mind. The results of this view, and of the inquiries which it suggested, are embodied in the following Treatise.

It is not generally permitted to an Author to prescribe the mode in which his production shall be judged; but there are two conditions which I may venture to require of those who shall undertake to estimate the merits of this performance. The first is, that no preconceived notion of the impossibility of its objects shall be permitted to interfere with that candour and impartiality which the investigation of Truth demands; the second is, that their judgment of the system as a whole shall not be founded either upon the examination of only

* See p. 42.

a part of it, or upon the measure of its conformity with any received system, considered as a standard of reference from which appeal is denied. It is in the general theorems which occupy the latter chapters of this work,—results to which there is no existing counterpart,—that the claims of the method, as a Calculus of Deductive Reasoning, are most fully set forth.

What may be the final estimate of the value of the system, I have neither the wish nor the right to anticipate. The estimation of a theory is not simply determined by its truth It also depends upon the importance of its subject, and the extent of its applications; beyond which something must still be left to the arbitrariness of human Opinion. If the utility of the application of Mathematical forms to the science of Logic were solely a question of Notation, I should be content to rest the defence of this attempt upon a principle which has been stated by an able living writer: "Whenever the nature of the subject permits the reasoning process to be without danger carried on mechanically, the language should be constructed on as mechanical principles as possible; while in the contrary case it should be so constructed, that there shall be the greatest possible obstacle to a mere mechanical use of it."* In one respect, the science of Logic differs from all others; the perfection of its method is chiefly valuable as an evidence of the speculative truth of its principles. To supersede the employment of common reason, or to subject it to the rigour of technical forms, would be the last desire of one who knows the value of that intellectual toil and warfare which imparts to the mind an athletic vigour, and teaches it to contend with difficulties and to rely upon itself in emergencies.

* Mill's *System of Logic, Ratiocinative and Inductive*, Vol. II. p. 292.

LINCOLN, *Oct.* 29, 1847.

MATHEMATICAL ANALYSIS OF LOGIC.

INTRODUCTION.

THEY who are acquainted with the present state of the theory
of Symbolical Algebra, are aware, that the validity of the
processes of analysis does not depend upon the interpretation
of the symbols which are employed, but solely upon the laws
of their combination. Every system of interpretation which
does not affect the truth of the relations supposed, is equally
admissible, and it is thus that the same process may, under
one scheme of interpretation, represent the solution of a question on the properties of numbers, under another, that of
a geometrical problem, and under a third, that of a problem
of dynamics or optics. This principle is indeed of fundamental
importance; and it may with safety be affirmed, that the recent
advances of pure analysis have been much assisted by the
influence which it has exerted in directing the current of
investigation.

But the full recognition of the consequences of this important
doctrine has been, in some measure, retarded by accidental
circumstances. It has happened in every known form of
analysis, that the elements to be determined have been conceived as measurable by comparison with some fixed standard.
The predominant idea has been that of magnitude, or more
strictly, of numerical ratio. The expression of magnitude, or

of operations upon magnitude, has been the express object for which the symbols of Analysis have been invented, and for which their laws have been investigated. Thus the abstractions of the modern Analysis, not less than the ostensive diagrams of the ancient Geometry, have encouraged the notion, that Mathematics are essentially, as well as actually, the Science of Magnitude.

The consideration of that view which has already been stated, as embodying the true principle of the Algebra of Symbols, would, however, lead us to infer that this conclusion is by no means necessary. If every existing interpretation is shewn to involve the idea of magnitude, it is only by induction that we can assert that no other interpretation is possible. And it may be doubted whether our experience is sufficient to render such an induction legitimate. The history of pure Analysis is, it may be said, too recent to permit us to set limits to the extent of its applications. Should we grant to the inference a high degree of probability, we might still, and with reason, maintain the sufficiency of the definition to which the principle already stated would lead us. We might justly assign it as the definitive character of a true Calculus, that it is a method resting upon the employment of Symbols, whose laws of combination are known and general, and whose results admit of a consistent interpretation. That to the existing forms of Analysis a quantitative interpretation is assigned, is the result of the circumstances by which those forms were determined, and is not to be construed into a universal condition of Analysis. It is upon the foundation of this general principle, that I purpose to establish the Calculus of Logic, and that I claim for it a place among the acknowledged forms of Mathematical Analysis, regardless that in its object and in its instruments it must at present stand alone.

That which renders Logic possible, is the existence in our minds of general notions,—our ability to conceive of a class, and to designate its individual members by a common name.

The theory of Logic is thus intimately connected with that of Language. A successful attempt to express logical propositions by symbols, the laws of whose combinations should be founded upon the laws of the mental processes which they represent, would, so far, be a step toward a philosophical language. But this is a view which we need not here follow into detail.* Assuming the notion of a class, we are able, from any conceivable collection of objects, to separate by a mental act, those which belong to the given class, and to contemplate them apart from the rest. Such, or a similar act of election, we may conceive to be repeated. The group of individuals left under consideration may be still further limited, by mentally selecting those among them which belong to some other recognised class, as well as to the one before contemplated. And this process may be repeated with other elements of distinction, until we arrive at an individual possessing all the distinctive characters which we have taken into account, and a member, at the same time, of every class which we have enumerated. It is in fact a method similar to this which we employ whenever, in common language, we accumulate descriptive epithets for the sake of more precise definition.

Now the several mental operations which in the above case we have supposed to be performed, are subject to peculiar laws. It is possible to assign relations among them, whether as respects the repetition of a given operation or the succession of different ones, or some other particular, which are never violated. It is, for example, true that the result of two successive acts is

---

* This view is well expressed in one of Blanco White's Letters :—" Logic is for the most part a collection of technical rules founded on classification. The Syllogism is nothing but a result of the classification of things, which the mind naturally and necessarily forms, in forming a language. All abstract terms are classifications ; or rather the labels of the classes which the mind has settled." —*Memoirs of the Rev. Joseph Blanco White*, vol. ii. p. 163. See also, for a very lucid introduction, Dr. Latham's *First Outlines of Logic applied to Language,* Becker's *German Grammar, &c.* Extreme Nominalists make Logic entirely dependent upon language. For the opposite view, see Cudworth's *Eternal and Immutable Morality*, Book vi. Chap. iii.

unaffected by the order in which they are performed ; and there
are at least two other laws which will be pointed out in the
proper place.   These will perhaps to some appear so obvious as
to be ranked among necessary truths, and so little important
as to be undeserving of special notice.   And probably they are
noticed for the first time in this Essay.   Yet it may with confidence be asserted, that if they were other than they are, the
entire mechanism of reasoning, nay the very laws and constitution of the human intellect, would be vitally changed.   A Logic
might indeed exist, but it would no longer be the Logic we
possess.

Such are the elementary laws upon the existence of which,
and upon their capability of exact symbolical expression, the
method of the following Essay is founded ; and it is presumed
that the object which it seeks to attain will be thought to
have been very fully accomplished.   Every logical proposition,
whether categorical or hypothetical, will be found to be capable
of exact and rigorous expression, and not only will the laws of
conversion and of syllogism be thence deducible, but the resolu-
tion of the most complex systems of propositions, the separation
of any proposed element, and the expression of its value in
terms of the remaining elements, with every subsidiary relation involved.   Every process will represent deduction, every
mathematical consequence will express a logical inference.   The
generality of the method will even permit us to express arbitrary operations of the intellect, and thus lead to the demonstration of general theorems in logic analogous, in no slight
degree, to the general theorems of ordinary mathematics.   No
inconsiderable part of the pleasure which we derive from the
application of analysis to the interpretation of external nature,
arises from the conceptions which it enables us to form of the
universality of the dominion of law.   The general formulæ to
which we are conducted seem to give to that element a visible
presence, and the multitude of particular cases to which they
apply, demonstrate the extent of its sway.   Even the symmetry

of their analytical expression may in no fanciful sense be deemed indicative of its harmony and its consistency. Now I do not presume to say to what extent the same sources of pleasure are opened in the following Essay. The measure of that extent may be left to the estimate of those who shall think the subject worthy of their study. But I may venture to assert that such occasions of intellectual gratification are not here wanting. The laws we have to examine are the laws of one of the most important of our mental faculties. The mathematics we have to construct are the mathematics of the human intellect. Nor are the form and character of the method, apart from all regard to its interpretation, undeserving of notice. There is even a remarkable exemplification, in its general theorems, of that species of excellence which consists in freedom from exception. And this is observed where, in the corresponding cases of the received mathematics, such a character is by no means apparent. The few who think that there is that in analysis which renders it deserving of attention for its own sake, may find it worth while to study it under a form in which every equation can be solved and every solution interpreted. Nor will it lessen the interest of this study to reflect that every peculiarity which they will notice in the form of the Calculus represents a corresponding feature in the constitution of their own minds.

It would be premature to speak of the value which this method may possess as an instrument of scientific investigation. I speak here with reference to the theory of reasoning, and to the principle of a true classification of the forms and cases of Logic considered as a Science.* The aim of these investigations was in the first instance confined to the expression of the received logic, and to the forms of the Aristotelian arrangement,

---

* "Strictly a Science"; also "an Art."—*Whately's Elements of Logic.* Indeed ought we not to regard all Art as applied Science ; unless we are willing, with "the multitude," to consider Art as "guessing and aiming well"?—*Plato, Philebus.*

but it soon became apparent that restrictions were thus introduced, which were purely arbitrary and had no foundation in
the nature of things. These were noted as they occurred, and
will be discussed in the proper place. When it became necessary to consider the subject of hypothetical propositions (in which
comparatively less has been done), and still more, when an
interpretation was demanded for the general theorems of the
Calculus, it was found to be imperative to dismiss all regard for
precedent and authority, and to interrogate the method itself for
an expression of the just limits of its application. Still, however, there was no special effort to arrive at novel results. But
among those which at the time of their discovery appeared to be
such, it may be proper to notice the following.

A logical proposition is, according to the method of this Essay,
expressible by an equation the form of which determines the
rules of conversion and of transformation, to which the given
proposition is subject. Thus the law of what logicians term
simple conversion, is determined by the fact, that the corresponding equations are symmetrical, that they are unaffected by
a mutual change of place, in those symbols which correspond
to the convertible classes. The received laws of conversion
were thus determined, and afterwards another system, which is
thought to be more elementary, and more general. See Chapter,
On the Conversion of Propositions.

The premises of a syllogism being expressed by equations, the
elimination of a common symbol between them leads to a third
equation which expresses the conclusion, this conclusion being
always the most general possible, whether Aristotelian or not.
Among the cases in which no inference was possible, it was
found, that there were two distinct forms of the final equation.
It was a considerable time before the explanation of this first
was discovered, but it was at length seen to depend upon the
presence or absence of a true medium of comparison between
the premises. The distinction which is thought to be new
is illustrated in the Chapter, *On Syllogisms.*

The nonexclusive character of the disjunctive conclusion of a hypothetical syllogism, is very clearly pointed out in the examples of this species of argument.

The class of logical problems illustrated in the chapter, *On the Solution of Elective Equations*, is conceived to be new : and it is believed that the method of that chapter affords the means of a perfect analysis of any conceivable system of propositions, an end toward which the rules for the conversion of a single categorical proposition are but the first step.

However, upon the originality of these or any of these views, I am conscious that I possess too slight an acquaintance with the literature of logical science, and especially with its older literature, to permit me to speak with confidence.

It may not be inappropriate, before concluding these observations, to offer a few remarks upon the general question of the use of symbolical language in the mathematics. Objections have lately been very strongly urged against this practice, on the ground, that by obviating the necessity of thought, and substituting a reference to general formulæ in the room of personal effort, it tends to weaken the reasoning faculties.

Now the question of the use of symbols may be considered in two distinct points of view. First, it may be considered with reference to the progress of scientific discovery, and secondly, with reference to its bearing upon the discipline of the intellect.

And with respect to the first view, it may be observed that as it is one fruit of an accomplished labour, that it sets us at liberty to engage in more arduous toils, so it is a necessary result of an advanced state of science, that we are permitted, and even called upon, to proceed to higher problems, than those which we before contemplated. The practical inference is obvious. If through the advancing power of scientific methods, we find that the pursuits on which we were once engaged, afford no longer a sufficiently ample field for intellectual effort, the remedy is, to proceed to higher inquiries, and, in new tracks, to seek for difficulties yet unsubdued. And such is, indeed, the actual law of scientific progress. We must be content, either to abandon the hope of further conquest, or to employ such aids of symbolical language, as are proper to the stage of progress, at which we have arrived. Nor need we fear to commit ourselves to such a course. We have not yet arrived so near to the boundaries of possible knowledge, as to suggest the apprehension, that scope will fail for the exercise of the inventive faculties.

In discussing the second, and scarcely less momentous question of the influence of the use of symbols upon the discipline of the intellect, an important distinction ought to be made. It is of most material consequence, whether those symbols are used with a full understanding of their meaning, with a perfect comprehension of that which renders their use lawful, and an ability to expand the abbreviated forms of reasoning which they induce, into their full syllogistic devolopment; or whether they are mere unsuggestive characters, the use of which is suffered to rest upon authority.

The answer which must be given to the question proposed, will differ according as the one or the other of these suppositions is admitted. In the former case an intellectual discipline of a high order is provided, an exercise not only of reason, but of the faculty of generalization. In the latter case there is no mental discipline whatever. It were perhaps the best security against the danger of an unreasoning reliance upon symbols, on the one hand, and a neglect of their just claims on the other, that each subject of applied mathematics should be treated in the spirit of the methods which were known at the time when the application was made, but in the best form which those methods have assumed. The order of attainment in the individual mind would thus bear some relation to the actual order of scientific discovery, and the more abstract methods of the higher analysis would be offered to such minds only, as were prepared to receive them.

The relation in which this Essay stands at once to Logic and

to Mathematics, may further justify some notice of the question
which has lately been revived, as to the relative value of the two
studies in a liberal education. One of the chief objections which
have been urged against the study of Mathematics in general, is
but another form of that which has been already considered with
respect to the use of symbols in particular. And it need not here
be further dwelt upon, than to notice, that if it avails anything,
it applies with an equal force against the study of Logic. The
canonical forms of the Aristotelian syllogism are really symbolical; only the symbols are less perfect of their kind than those
of mathematics. If they are employed to test the validity of an
argument, they as truly supersede the exercise of reason, as does
a reference to a formula of analysis. Whether men do, in the
present day, make this use of the Aristotelian canons, except as
a special illustration of the rules of Logic, may be doubted; yet
it cannot be questioned that when the authority of Aristotle was
dominant in the schools of Europe, such applications were habit-
ually made. And our argument only requires the admission,
that the case is possible.

But the question before us has been argued upon higher
grounds. Regarding Logic as a branch of Philosophy, and de-
fining Philosophy as the "science of a real existence," and "the
research of causes," and assigning as its *main* business the inves-
tigation of the " why, ($\tau \grave{o}$ $\delta\iota o\tau\iota$)," while Mathematics display
only the " that, ($\tau \grave{o}$ $\acute{o}\tau\iota$)," Sir W. Hamilton has contended,
not simply, that the superiority rests with the study of Logic,
but that the study of Mathematics is at once dangerous and use-
less.* The pursuits of the mathematician " have not only not
trained him to that acute scent, to that delicate, almost instinc-
tive, tact which, in the twilight of probability, the search and
discrimination of its finer facts demand; they have gone to cloud
his vision, to indurate his touch, to all but the blazing light, the
iron chain of demonstration, and left him out of the narrow con-
fines of his science, to a passive *credulity* in any premises, or to

* *Edinburgh Review*, vol. LXII. p. 409, and *Letter to A. De Morgan, Esq.*

an absolute *incredulity* in all." In support of these and of other charges, both argument and copious authority are adduced.* I shall not attempt a complete discussion of the topics which are suggested by these remarks. My object is not controversy, and the observations which follow are offered not in the spirit of antagonism, but in the hope of contributing to the formation of just views upon an important subject. Of Sir W. Hamilton it is impossible to speak otherwise than with that respect which is due to genius and learning.

Philosophy is then described as the *science of a real existence* and *the research of causes.* And that no doubt may rest upon the meaning of the word *cause,* it is further said, that philosophy " mainly investigates the *why.*" These definitions are common among the ancient writers. Thus Seneca, one of Sir W. Hamilton's authorities, *Epistle* LXXXVIII., " The philosopher seeks and knows the *causes* of natural things, of which the mathematician searches out and computes the numbers and the measures." It may be remarked, in passing, that in whatever degree the belief has prevailed, that the business of philosophy is immediately with *causes;* in the same degree has every science whose object is the investigation of *laws,* been lightly esteemed. Thus the Epistle to which we have referred, bestows, by contrast with Philosophy, a separate condemnation on Music and Grammar, on Mathematics and Astronomy, although it is that of Mathematics only that Sir W. Hamilton has quoted.

Now we might take our stand upon the conviction of many thoughtful and reflective minds, that in the extent of the meaning above stated, Philosophy is impossible. The business of true Science, they conclude, is with laws and phenomena. The nature of Being, the mode of the operation of Cause, the *why,*

---

* The arguments are in general better than the authorities. Many writers quoted in condemnation of mathematics (Aristo, Seneca, Jerome, Augustine, Cornelius Agrippa, &c.) have borne a no less explicit testimony against other sciences, nor least of all, against that of logic. The treatise of the last named writer *De Vanitate Scientiarum,* must surely have been referred to by mistake.— *Vide* cap. CII.

they hold to be beyond the reach of our intelligence. But we do not require the vantage-ground of this position; nor is it doubted that whether the aim of Philosophy is attainable or not, the desire which impels us to the attempt is an instinct of our higher nature. Let it be granted that the problem which has baffled the efforts of ages, is not a hopeless one; that the " science of a real existence," and " the research of causes," " that kernel" for which " Philosophy is still militant," do not transcend the limits of the human intellect. I am then compelled to assert, that according to this view of the nature of Philosophy, *Logic forms no part of it.* On the principle of a true classification, we ought no longer to associate Logic and Metaphysics, but Logic and Mathematics.

Should any one after what has been said, entertain a doubt upon this point, I must refer him to the evidence which will be afforded in the following Essay. He will there see Logic resting like Geometry upon axiomatic truths, and its theorems constructed upon that general doctrine of symbols, which constitutes the foundation of the recognised Analysis. In the Logic of Aristotle he will be led to view a collection of the formulæ of the science, expressed by another, but, (it is thought) less perfect scheme of symbols. I feel bound to contend for the absolute exactness of this parallel. It is no escape from the conclusion to which it points to assert, that Logic not only constructs a science, but also inquires into the origin and the nature of its own principles,—a distinction which is denied to Mathematics. " It is wholly beyond the domain of mathematicians," it is said, " to inquire into the origin and nature of their principles."— *Review,* page 415. But upon what ground can such a distinction be maintained? What definition of the term Science will be found sufficiently arbitrary to allow such differences?

The application of this conclusion to the question before us is clear and decisive. The mental discipline which is afforded by the study of Logic, *as an exact science,* is, in species, the same as that afforded by the study of Analysis.

Is it then contended that either Logic or Mathematics can supply a perfect discipline to the Intellect? The most careful and unprejudiced examination of this question leads me to doubt whether such a position can be maintained. The exclusive claims of either must, I believe, be abandoned, nor can any others, partaking of a like exclusive character, be admitted in their room. It is an important observation, which has more than once been made, that it is one thing to arrive at correct premises, and another thing to deduce logical conclusions, and that the business of life depends more upon the former than upon the latter. The study of the exact sciences may teach us the one, and it may give us some general preparation of knowledge and of practice for the attainment of the other, but it is to the union of thought with action, in the field of Practical Logic, the arena of Human Life, that we are to look for its fuller and more perfect accomplishment.

I desire here to express my conviction, that with the advance of our knowledge of all true science, an ever-increasing harmony will be found to prevail among its separate branches. The view which leads to the rejection of one, ought, if consistent, to lead to the rejection of others. And indeed many of the authorities which have been quoted against the study of Mathematics, are even more explicit in their condemnation of Logic. " Natural science," says the Chian Aristo, "is above us, Logical science does not concern us." When such conclusions are founded (as they often are) upon a deep conviction of the preeminent value and importance of the study of Morals, we admit the premises, but must demur to the inference. For it has been well said by an ancient writer, that it is the " characteristic of the liberal sciences, not that they conduct us to Virtue, but that they prepare us for Virtue;" and Melancthon's sentiment, " abeunt studia in mores," has passed into a proverb. Moreover, there is a common ground upon which all sincere votaries of truth may meet, exchanging with each other the language of Flamsteed's appeal to Newton, " The works of the Eternal Providence will be better understood through your labors and mine."

FIRST PRINCIPLES.

Let us employ the symbol 1, or unity, to represent the Universe, and let us understand it as comprehending every conceivable class of objects whether actually existing or not, it being premised that the same individual may be found in more than one class, inasmuch as it may possess more than one quality in common with other individuals. Let us employ the letters X, Y, Z, to represent the individual members of classes, X applying to every member of one class, as members of that particular class, and Y to every member of another class as members of such class, and so on, according to the received language of treatises on Logic.

Further let us conceive a class of symbols x, y, z, possessed of the following character.

The symbol x operating upon any subject comprehending individuals or classes, shall be supposed to select from that subject all the Xs which it contains. In like manner the symbol y, operating upon any subject, shall be supposed to select from it all individuals of the class Y which are comprised in it, and so on.

When no subject is expressed, we shall suppose 1 (the Universe) to be the subject understood, so that we shall have

$$x = x \quad (1),$$

the meaning of either term being the selection from the Universe of all the Xs which it contains, and the result of the operation being in common language, the class X, *i. e.* the class of which each member is an X.

From these premises it will follow, that the product *xy* will represent, in succession, the selection of the class Y, and the selection from the class Y of such individuals of the class X as are contained in it, the result being the class whose members are both Xs and Ys. And in like manner the product *xyz* will represent a compound operation of which the successive elements are the selection of the class Z, the selection from it of such individuals of the class Y as are contained in it, and the selection from the result thus obtained of all the individuals of the class X which it contains, the final result being the class common to X, Y, and Z.

From the nature of the operation which the symbols *x, y, z,* are conceived to represent, we shall designate them as elective symbols. An expression in which they are involved will be called an elective function, and an equation of which the members are elective functions, will be termed an elective equation.

It will not be necessary that we should here enter into the analysis of that mental operation which we have represented by the elective symbol. It is not an act of Abstraction according to the common acceptation of that term, because we never lose sight of the concrete, but it may probably be referred to an exercise of the faculties of Comparison and Attention. Our present concern is rather with the laws of combination and of succession, by which its results are governed, and of these it will suffice to notice the following.

1st. The result of an act of election is independent of the grouping or classification of the subject.

Thus it is indifferent whether from a group of objects considered as a whole, we select the class X, or whether we divide the group into two parts, select the Xs from them separately, and then connect the results in one aggregate conception.

We may express this law mathematically by the equation

$$x \, (u + v) = xu + xv,$$

$u + v$ representing the undivided subject, and u and v the component parts of it.

2nd. It is indifferent in what order two successive acts of election are performed.

Whether from the class of animals we select sheep, and from the sheep those which are horned, or whether from the class of animals we select the horned, and from these such as are sheep, the result is unaffected. In either case we arrive at the class *horned sheep*.

The symbolical expression of this law is

$$xy = yx.$$

3rd. The result of a given act of abstraction performed twice, or any number of times in succession, is the result of the same act performed once.

If from a group of objects we select the Xs, we obtain a class of which all the members are Xs. If we repeat the operation on this class no further change will ensue: in selecting the Xs we take the whole. Thus we have

$$xx = x,$$

or
$$x^2 = x;$$

and supposing the same operation to be n times performed, we have
$$x^n = x,$$

which is the mathematical expression of the law above stated.*

The laws we have established under the symbolical forms

$$x (u + v) = xu + xv \dots\dots\dots\dots\dots (1),$$
$$xy = yx \dots\dots\dots\dots\dots\dots (2),$$
$$x^n = x \dots\dots\dots\dots\dots\dots (3),$$

---

* The office of the elective symbol x, is to select individuals comprehended in the class X. Let the class X be supposed to embrace the universe; then, whatever the class Y may be, we have
$$xy = y.$$
The office which x performs is now equivalent to the symbol $+$, in one at least of its interpretations, and the index law (3) gives
$$+^n = +,$$
which is the known property of that symbol.

are sufficient for the basis of a Calculus. From the first of these, it appears that elective symbols are *distributive*, from the second that they are *commutative;* properties which they possess in common with symbols of *quantity,* and in virtue of which, all the processes of common algebra are applicable to the present system. The one and sufficient axiom involved in this application is that equivalent operations performed upon equivalent subjects produce equivalent results.*

The third law (3) we shall denominate the index law. It is peculiar to elective symbols, and will be found of great importance in enabling us to reduce our results to forms meet for interpretation.

From the circumstance that the processes of algebra may be applied to the present system, it is not to be inferred that the interpretation of an elective equation will be unaffected by such processes. The expression of a truth cannot be negatived by

* It is generally asserted by writers on Logic, that all reasoning ultimately depends on an application of the dictum of Aristotle, *de omni et nullo.* " Whatever is predicated universally of any class of things, may be predicated in like manner of any thing comprehended in that class." But it is agreed that this dictum is not immediately applicable in all cases, and that in a majority of instances, a certain previous process of reduction is necessary. What are the elements involved in that process of reduction? Clearly they are as much a part of general reasoning as the dictum itself.

Another mode of considering the subject resolves all reasoning into an application of one or other of the following canons, viz.

1. If two terms agree with one and the same third, they agree with each other.

2. If one term agrees, and another disagrees, with one and the same third, these two disagree with each other.

But the application of these canons depends on mental acts equivalent to those which are involved in the before-named process of reduction. We have to select individuals from classes, to convert propositions, &c., before we can avail ourselves of their guidance. Any account of the process of reasoning is insufficient, which does not represent, as well the laws of the operation which the mind performs in that process, as the primary truths which it recognises and applies.

It is presumed that the laws in question are adequately represented by the fundamental equations of the present Calculus. The proof of this will be found in its capability of expressing propositions, and of exhibiting in the results of its processes, every result that may be arrived at by ordinary reasoning.

a legitimate operation, but it may be limited. The equation $y = z$ implies that the classes Y and Z are equivalent, member for member. Multiply it by a factor x, and we have

$$xy = xz,$$

which expresses that the individuals which are common to the classes X and Y are also common to X and Z, and *vice versâ*. This is a perfectly legitimate inference, but the fact which it declares is a less general one than was asserted in the original proposition.

OF EXPRESSION AND INTERPRETATION.

A Proposition is a sentence which either affirms or denies, as, All men are mortal, No creature is independent.

A Proposition has necessarily two terms, as *men*, *mortal;* the former of which, or the one spoken of, is called the subject; the latter, or that which is affirmed or denied of the subject, the predicate. These are connected together by the copula *is*, or *is not*, or by some other modification of the substantive verb.

The substantive verb is the only verb recognised in Logic; all others are resolvable by means of the verb *to be* and a participle or adjective, *e. g.* "The Romans conquered"; the word conquered is both copula and predicate, being equivalent to "were (copula) victorious" (predicate).

A Proposition must either be affirmative or negative, and must be also either universal or particular. Thus we reckon in all, four kinds of pure categorical Propositions.

1st. Universal-affirmative, usually represented by A,

Ex. All Xs are Ys.

2nd. Universal-negative, usually represented by E,

Ex. No Xs are Ys.

3rd. Particular-affirmative, usually represented by I,

Ex. Some Xs are Ys.

4th. Particular-negative, usually represented by O,*

Ex. Some Xs are not Ys.

1. To express the class, not-X, that is, the class including all individuals that are not Xs.

The class X and the class not-X together make the Universe. But the Universe is 1, and the class X is determined by the symbol x, therefore the class not-X will be determined by the symbol $1 - x$.

* The above is taken, with little variation, from the Treatises of Aldrich and Whately.

Hence the office of the symbol $1 - x$ attached to a given subject will be, to select from it all the not-Xs which it contains.

And in like manner, as the product xy expresses the entire class whose members are both Xs and Ys, the symbol $y(1 - x)$ will represent the class whose members are Ys but not Xs, and the symbol $(1 - x)(1 - y)$ the entire class whose members are neither Xs nor Ys.

2. To express the Proposition, All Xs are Ys.

As all the Xs which exist are found in the class Y, it is obvious that to select out of the Universe all Ys, and from these to select all Xs, is the same as to select at once from the Universe all Xs.

Hence $$xy = x,$$

or $$x(1 - y) = 0, \quad (4).$$

3. To express the Proposition, No Xs are Ys.

To assert that no Xs are Ys, is the same as to assert that there are no terms common to the classes X and Y. Now all individuals common to those classes are represented by xy. Hence the Proposition that No Xs are Ys, is represented by the equation

$$xy = 0, \quad (5).$$

4. To express the Proposition, Some Xs are Ys.

If some Xs are Ys, there are some terms common to the classes X and Y. Let those terms constitute a separate class V, to which there shall correspond a separate elective symbol v, then

$$v = xy, \quad (6).$$

And as v includes all terms common to the classes X and Y, we can indifferently interpret it, as Some Xs, or Some Ys.

5. To express the Proposition, Some Xs are not Ys.

In the last equation write $1 - y$ for y, and we have

$$v = x\,(1 - y), \quad (7),$$

the interpretation of v being indifferently Some Xs or Some not-Ys.

The above equations involve the complete theory of categorical Propositions, and so far as respects the employment of analysis for the deduction of logical inferences, nothing more can be desired. But it may be satisfactory to notice some particular forms deducible from the third and fourth equations, and susceptible of similar application.

If we multiply the equation (6) by x, we have

$$vx = x^2y = xy \text{ by (3).}$$

Comparing with (6), we find

$$v = vx,$$

or $$v\,(1 - x) = 0, \quad (8).$$

And multiplying (6) by y, and reducing in a similar manner, we have

$$v = vy,$$

or $$v\,(1 - y) = 0, \quad (9).$$

Comparing (8) and (9),

$$vx = vy = v, \quad (10).$$

And further comparing (8) and (9) with (4), we have as the equivalent of this system of equations the Propositions

$$\left.\begin{array}{l} \text{All Vs are Xs} \\ \text{All Vs are Ys} \end{array}\right\}$$

The system (10) might be used to replace (6), or the single equation

$$vx = vy, \quad (11),$$

might be used, assigning to vx the interpretation, Some Xs, and to vy the interpretation, Some Ys. But it will be observed that this system does not express quite so much as the single equation (6), from which it is derived. Both, indeed, express the Proposition, Some Xs are Ys, but the system (10) does not imply that the class V includes *all* the terms that are common to X and Y.

In like manner, from the equation (7) which expresses the Proposition Some Xs are not Ys, we may deduce the system

$$vx = v(1 - y) = v, \quad (12),$$

in which the interpretation of $v(1 - y)$ is Some not-Ys. Since in this case $vy = 0$, we must of course be careful not to interpret vy as Some Ys.

If we multiply the first equation of the system (12), viz.

$$vx = v(1 - y),$$

by y, we have

$$vxy = vy(1 - y);$$

$$\therefore vxy = 0, \quad (13),$$

which is a form that will occasionally present itself. It is not necessary to revert to the primitive equation in order to interpret this, for the condition that vx represents Some Xs, shews us by virtue of (5), that its import will be

Some Xs are not Ys,

the subject comprising *all* the Xs that are found in the class V.

Universally in these cases, difference of form implies a difference of interpretation with respect to the auxiliary symbol v, and each form is interpretable by itself.

Further, these differences do not introduce into the Calculus a needless perplexity. It will hereafter be seen that they give a precision and a definiteness to its conclusions, which could not otherwise be secured.

Finally, we may remark that all the equations by which particular truths are expressed, are deducible from any one general equation, expressing any one general Proposition, from which those particular Propositions are necessary deductions.

This has been partially shewn already, but it is much more fully exemplified in the following scheme.

The general equation $\quad x = y,$

implies that the classes X and Y are equivalent, member for member; that every individual belonging to the one, belongs to the other also. Multiply the equation by x, and we have

$$x^2 = xy;$$
$$\therefore x = xy,$$

which implies, by (4), that all Xs are Ys. Multiply the same equation by y, and we have in like manner

$$y = xy;$$

the import of which is, that all Ys are Xs. Take either of these equations, the latter for instance, and writing it under the form

$$(1 - x)\, y = 0,$$

we may regard it as an equation in which y, an unknown quantity, is sought to be expressed in terms of x. Now it will be shewn when we come to treat of the Solution of Elective Equations (and the result may here be verified by substitution) that the most general solution of this equation is

$$y = vx,$$

which implies that All Ys are Xs, and that Some Xs are Ys. Multiply by x, and we have

$$vy = vx,$$

which indifferently implies that some Ys are Xs and some Xs are Ys, being the particular form at which we before arrived.

For convenience of reference the above and some other results have been classified in the annexed Table, the first column of which contains propositions, the second equations, and the third the conditions of final interpretation. It is to be observed, that the auxiliary equations which are given in this column are not independent: they are implied either in the equations of the second column, or in the condition for the interpretation of v. But it has been thought better to write them separately, for greater ease and convenience. And it is further to be borne in mind, that although three different forms are given for the expression of each of the *particular* propositions, everything is really included in the first form.

TABLE.

The class X	x	
The class not-X	$1 - x$	
All Xs are Ys⎱ All Ys are Xs⎰	$x = y$	
All Xs are Ys	$x (1 - y) = 0$	
No Xs are Ys	$xy = 0$	
All Ys are Xs ⎱ Some Xs are Ys⎰ $y = vx$		$vx = $ some Xs $v (1 - x) = 0.$
No Ys are Xs ⎱ Some not-Xs are Ys⎰ $y = v (1 - x)$		$v (1 - x) = $ some not-Xs $vx = 0.$
Some Xs are Ys	$\begin{cases} v = xy \\ \text{or } vx = vy \\ \text{or } vx (1 - y) = 0 \end{cases}$	$v = $ some Xs or some Ys $vx = $ some Xs, $vy = $ some Ys $v (1 - x) = 0, v (1 - y) = 0.$
Some Xs are not Ys	$\begin{cases} v = x (1 - y) \\ \text{or } vx = v (1 - y) \\ \text{or } vxy = 0 \end{cases}$	$v = $ some Xs, or some not-Ys $vx = $ some Xs, $v (1 - y) = $ some not-Ys $v (1 - x) = 0, vy = 0.$

OF THE CONVERSION OF PROPOSITIONS.

A Proposition is said to be converted when its terms are transposed; when nothing more is done, this is called simple conversion; *e.g.*

> No virtuous man is a tyrant, *is converted into*
> No tyrant is a virtuous man.

Logicians also recognise conversion *per accidens*, or by limitation, *e.g.*

> All birds are animals, *is converted into*
> Some animals are birds.

And conversion by *contraposition* or *negation*, as

> Every poet is a man of genius, *converted into*
> He who is not a man of genius is not a poet.

In one of these three ways every Proposition may be illatively converted, viz. E and I simply, A and O by negation, A and E by limitation.

The primary canonical forms already determined for the expression of Propositions, are

All Xs are Ys, $\qquad x(1-y)=0,\qquad$ ....A.

No Xs are Ys, $\qquad\qquad xy=0,\qquad$ ....E.

Some Xs are Ys, $\qquad\qquad v=xy,\qquad$ .... I.

Some Xs are not Ys, $\qquad v=x(1-y)$ ....O.

On examining these, we perceive that E and I are symmetrical with respect to x and y, so that x being changed into y, and y into x, the equations remain unchanged. Hence E and I may be interpreted into

No Ys are Xs,

Some Ys are Xs,

respectively. Thus we have the known rule of the Logicians, that particular affirmative and universal negative Propositions admit of simple conversion.

The equations A and O may be written in the forms

$$(1 - y) \left\{ 1 - (1 - x) \right\} = 0,$$

$$v = (1 - y) \left\{ 1 - (1 - x) \right\}.$$

Now these are precisely the forms which we should have obtained if we had in those equations changed x into $1 - y$, and y into $1 - x$, which would have represented the changing in the original Propositions of the Xs into not-Ys, and the Ys into not-Xs, the resulting Propositions being

All not-Ys are not-Xs,

Some not-Ys are not not-Xs (a).

Or we may, by simply inverting the order of the factors in the second member of O, and writing it in the form

$$v = (1 - y) \, x,$$

interpret it by I into

Some not-Ys are Xs,

which is really another form of (a). Hence follows the rule, that universal affirmative and particular negative Propositions admit of negative conversion, or, as it is also termed, conversion by contraposition.

The equations A and E, written in the forms

$$(1 - y) \, x = 0,$$

$$yx = 0,$$

give on solution the respective forms

$$x = vy,$$

$$x = v \, (1 - y),$$

the correctness of which may be shewn by substituting these values of x in the equations to which they belong, and observing that those equations are satisfied quite independently of the nature of the symbol v. The first solution may be interpreted into

Some Ys are Xs,

and the second into

Some not-Ys are Xs.

From which it appears that universal-affirmative, and universal-negative Propositions are convertible by limitation, or, as it has been termed, *per accidens.*

The above are the laws of Conversion recognized by Abp. Whately. Writers differ however as to the admissibility of negative conversion. The question depends on whether we will consent to use such terms as not-X, not-Y. Agreeing with those who think that such terms ought to be admitted, even although they change the *kind* of the Proposition, I am constrained to observe that the present classification of them is faulty and defective. Thus the conversion of No Xs are Ys, into All Ys are not-Xs, though perfectly legitimate, is not recognised in the above scheme. It may therefore be proper to examine the subject somewhat more fully.

Should we endeavour, from the system of equations we have obtained, to deduce the laws not only of the conversion, but also of the general transformation of propositions, we should be led to recognise the following distinct elements, each connected with a distinct mathematical process.

1st. The negation of a term, *i. e.* the changing of X into not-X, or not-X into X.

2nd. The translation of a Proposition from one *kind* to another, as if we should change

All Xs are Ys into Some Xs are Ys   A into I,

which would be lawful; or

All Xs are Ys into No Xs are Y.   A into E,

which would be unlawful.

3rd. The simple conversion of a Proposition.

The conditions in obedience to which these processes may lawfully be performed, may be deduced from the equations by which Propositions are expressed.

We have

$$\text{All Xs are Ys}\ldots\ldots x\,(1-y)=0. \qquad \text{A,}$$
$$\text{No Xs are Ys}\ldots\ldots\ldots xy=0. \qquad \text{E.}$$

Write E in the form

$$x \left\{ 1 - (1 - y) \right\} = 0,$$

and it is interpretable by A into

All Xs are not-Ys,

so that we may change

No Xs are Ys into All Xs are not-Ys.

In like manner A interpreted by E gives

No Xs are not-Ys,

so that we may change

All Xs are Ys into No Xs are not-Ys.

From these cases we have the following Rule: A universal-affirmative Proposition is convertible into a universal-negative, and, *vice versâ*, by negation of the predicate.

Again, we have

Some Xs are Ys. . . . .$v = xy$,

Some Xs are not Ys . . . .$v = x(1 - y)$.

These equations only differ from those last considered by the presence of the term v. The same reasoning therefore applies, and we have the Rule—

A particular-affirmative proposition is convertible into a particular-negative, and *vice versâ*, by negation of the predicate.

Assuming the universal Propositions

All Xs are Ys. . . . . .$x(1 - y) = 0$,

No Xs are Ys . . . . $xy = 0$.

Multiplying by v, we find

$$vx(1 - y) = 0,$$

$$vxy = 0,$$

which are interpretable into

Some Xs are Ys. . . . . . .I,

Some Xs are not Ys. . . .O.

Hence a universal-affirmative is convertible into a particular-affirmative, and a universal-negative into a particular-negative without negation of subject or predicate.

Combining the above with the already proved rule of simple conversion, we arrive at the following system of independent laws of transformation.

1st. An affirmative Proposition may be changed into its corresponding negative (A into E, or I into O), and *vice versa,* by negation of the predicate.

2nd. A universal Proposition may be changed into its corresponding particular Proposition, (A into I, or E into O).

3rd. In a particular-affirmative, or universal-negative Proposition, the terms may be mutually converted.

Wherein negation of a term is the changing of X into not-X, and *vice versâ,* and is not to be understood as affecting the *kind* of the Proposition.

Every lawful transformation is reducible to the above rules. Thus we have

> All Xs are Ys,
>
> No Xs are not-Ys   by 1st rule,
>
> No not-Ys are Xs   by 3rd rule,
>
> All not-Ys are not-Xs by 1st rule,

which is an example of *negative conversion.* Again,

> No Xs are Ys,
>
> No Ys are Xs   3rd rule,
>
> All Ys are not-Xs   1st rule,

which is the case already deduced.

OF SYLLOGISMS.

A Syllogism consists of three Propositions, the last of which, called the conclusion, is a logical consequence of the two former, called the premises; e.g.

$$\textit{Premises,} \quad \begin{cases} \text{All Ys are Xs.} \\ \text{All Zs are Ys.} \end{cases}$$

Conclusion, All Zs are Xs.

Every syllogism has three and only three terms, whereof that which is the subject of the conclusion is called the *minor* term, the predicate of the conclusion, the *major* term, and the remaining term common to both premises, the middle term. Thus, in ths above formula, Z is the minor term, X the major term, Y the middle term.

The figure of a syllogism consists in the situation of the middle term with respect to the terms of the conclusion. The varieties of figure are exhibited in the annexed scheme.

1st Fig.	2nd Fig.	3rd Fig.	4th Fig.
YX	XY	YX	XY
ZY	ZY	YZ	YZ
ZX	ZX	ZX	ZX

When we designate the three propositions of a syllogism by their usual symbols (A, E, I, O), and in their actual order, we are said to determine the mood of the syllogism. Thus the syllogism given above, by way of illustration, belongs to the mood AAA in the first figure.

The moods of all syllogisms commonly received as valid, are represented by the vowels in the following mnemonic verses.

Fig. 1.—bArbArA, cElArEnt, dArII, fErIO que prioris.

Fig. 2.—cEsArE, cAmEstrEs, fEstIno, bArOkO, secundæ.

Fig. 3.—Tertia dArAptI, dIsAmIs, dAtIsI, fElAptOn,
   bOkArdO, fErIsO, habet : quarta insuper addit.

Fig 4.—brAmAntIp, cAmEnEs, dImArIs, fEsapO, frEsIsOn.

THE equation by which we express any Proposition concerning the classes X and Y, is an equation between the symbols x and y, and the equation by which we express any

Proposition concerning the classes Y and Z, is an equation between the symbols y and z. If from two such equations we eliminate y, the result, if it do not vanish, will be an equation between x and z, and will be interpretable into a Proposition concerning the classes X and Z. And it will then constitute the third member, or Conclusion, of a Syllogism, of which the two given Propositions are the premises.

The result of the elimination of y from the equations

$$ay + b = 0,$$
$$a'y + b' = 0, \quad (14),$$

is the equation $\qquad ab' - a'b = 0, \quad (15)$.

Now the equations of Propositions being of the first order with reference to each of the variables involved, all the cases of elimination which we shall have to consider, will be reducible to the above case, the constants a, b, a', b', being replaced by functions of x, z, and the auxiliary symbol v.

As to the choice of equations for the expression of our premises, the only restriction is, that the equations must not *both* be of the form $ay = 0$, for in such cases elimination would be impossible. When both equations are of this form, it is necessary to solve one of them, and it is indifferent which we choose for this purpose. If that which we select is of the form $xy = 0$, its solution is

$$y = v(1 - x), \quad (16),$$

if of the form $(1 - x)y = 0$, the solution will be

$$y = vx, \quad (17),$$

and these are the only cases which can arise. The reason of this exception will appear in the sequel.

For the sake of uniformity we shall, in the expression of particular propositions, confine ourselves to the forms

$$vx = vy, \qquad \text{Some Xs are Ys,}$$
$$vx = v(1 - y), \qquad \text{Some Xs are not Ys,}$$

These have a closer analogy with (16) and (17), than the other forms which might be used.

Between the forms about to be developed, and the Aristotelian canons, some points of difference will occasionally be observed, of which it may be proper to forewarn the reader.

To the right understanding of these it is proper to remark, that the essential structure of a Syllogism is, in some measure, arbitrary. Supposing the order of the premises to be fixed, and the distinction of the major and the minor term to be thereby determined, it is purely a matter of choice which of the two shall have precedence in the Conclusion. Logicians have settled this question in favour of the minor term, but it is clear, that this is a convention. Had it been agreed that the major term should have the first place in the conclusion, a logical scheme might have been constructed, less convenient in some cases than the existing one, but superior in others. What it lost in *barbara*, it would gain in *bramantip*. Convenience is *perhaps* in favour of the adopted arrangement,* but it is to be remembered that it is *merely* an arrangement.

Now the method we shall exhibit, not having reference to one scheme of arrangement more than to another, will always give the more general conclusion, regard being paid only to its abstract lawfulness, considered as a result of pure reasoning. And therefore we shall sometimes have presented to us the spectacle of conclusions, which a logician would pronounce informal, but never of such as a reasoning being would account false.

The Aristotelian canons, however, beside restricting the *order* of the terms of a conclusion, limit their nature also;—and this limitation is of more consequence than the former. We may, by a change of figure, replace the particular conclusion

---

* The contrary view was maintained by Hobbes. The question is very fairly discussed in Hallam's *Introduction to the Literature of Europe*, vol. III. p. 309. In the rhetorical use of Syllogism, the advantage appears to rest with the rejected form.

of *bramantip*, by the general conclusion of *barbara;* but we cannot thus reduce to rule such inferences, as

Some not-Xs are not Ys.

Yet there are cases in which such inferences may lawfully be drawn, and in unrestricted argument they are of frequent occurrence. Now if an inference of this, or of any other kind, is lawful in itself, it will be exhibited in the results of our method.

We may by restricting the canon of interpretation confine our expressed results within the limits of the scholastic logic; but this would only be to restrict ourselves to the use of a part of the conclusions to which our analysis entitles us.

The classification we shall adopt will be purely mathematical, and we shall afterwards consider the logical arrangement to which it corresponds. It will be sufficient, for reference, to name the premises and the Figure in which they are found.

CLASS 1st.—Forms in which v does not enter.

Those which admit of an inference are AA, EA, Fig. 1; AE, EA, Fig. 2; AA, AE, Fig. 4.

Ex. AA, Fig. 1, and, by mutation of premises (change of order), AA, Fig. 4.

All Ys are Xs, $\qquad y(1-x)=0, \qquad$ or $(1-x)y=0,$

All Zs are Ys, $\qquad z(1-y)=0, \qquad$ or $\quad zy-z=0.$

Eliminating y by (15) we have

$$z(1-x)=0,$$

$$\therefore \text{ All Zs are Xs.}$$

A convenient mode of effecting the elimination, is to write the equation of the premises, so that y shall appear only as a factor of one member in the first equation, and only as a factor of the opposite member in the second equation, and then to multiply the equations, omitting the y. This method we shall adopt.

Ex. AE, Fig. 2, and, by mutation of premises, EA, Fig, 2.

All Xs are Ys,	$x(1 - y) = 0,$	or	$x = xy$
No Zs are Ys,	$zy = 0,$		$zy = 0$

$$zx = 0$$

$$\therefore \text{ No Zs are Xs.}$$

The only case in which there is no inference is AA, Fig. 2,

All Xs are Ys,	$x(1 - y) = 0,$	$x = xy$
All Zs are Ys,	$z(1 - y) = 0,$	$zy = z$

$$xz = xz$$

$$\therefore \quad 0 = 0.$$

CLASS 2nd.—When v is introduced by the solution of an equation.

The lawful cases directly or indirectly* determinable by the Aristotelian Rules are AE, Fig. 1; AA, AE, EA, Fig. 3; EA, Fig. 4.

The lawful cases not so determinable, are EE, Fig. 1; EE, Fig. 2; EE, Fig. 3; EE, Fig. 4.

Ex. AE, Fig. 1, and, by mutation of premises, EA, Fig. 4.

All Ys are Xs,	$y(1 - x) = 0,$	$y = vx$ (a)
No Zs are Ys,	$zy = 0,$	$0 = zy$

$$0 = vzx$$

$$\therefore \text{ Some Xs are not Zs.}$$

The reason why we cannot interpret $vzx = 0$ into Some Zs are not-Xs, is that by the very terms of the first equation (a) the interpretation of vx is fixed, as Some Xs; v is regarded as the representative of Some, only with reference to the class X.

---

* We say *directly* or *indirectly*, mutation or conversion of premises being in some instances required. Thus, AE (fig. 1) is resolvable by Fesapo (fig. 4), or by Ferio (fig. 1). Aristotle and his followers rejected the fourth figure as only a modification of the first, but this being a mere question of form, either scheme may be termed Aristotelian.

For the reason of our employing a solution of one of the primitive equations, see the remarks on (16) and (17). Had we solved the second equation instead of the first, we should have had

$$(1 - x)\, y = 0,$$
$$v\,(1 - z) = y, \quad (a),$$
$$v\,(1 - z)\,(1 - x) = 0, \quad (b),$$
$$\therefore \text{ Some not-Zs are Xs.}$$

Here it is to be observed, that the second equation (a) fixes the meaning of $v\,(1 - z)$, as Some not-Zs. The full meaning of the result (b) is, that all the not-Zs which are found in the class Y are found in the class X, and it is evident that this could not have been expressed in any other way.

Ex. 2. AA, Fig. 3.

All Ys are Xs,	$y\,(1 - x) = 0,$	$y = vx$
All Ys are Zs,	$y\,(1 - z) = 0,$	$0 = y\,(1 - z)$

$$0 = vx\,(1 - z)$$
$$\therefore \text{ Some Xs are Zs.}$$

Had we solved the second equation, we should have had as our result, Some Zs are Xs. The form of the final equation particularizes what Xs or what Zs are referred to, and this remark is general.

The following, EE, Fig. 1, and, by mutation, EE, Fig. 4, is an example of a lawful case not determinable by the Aristotelian Rules.

No Ys are Xs,	$xy = 0,$	$0 = xy$
No Zs are Ys,	$zy = 0,$	$y = v\,(1 - z)$

$$0 = v\,(1 - z)\, x$$
$$\therefore \text{ Some not-Zs are not Xs.}$$

CLASS 3rd.—When v is met with in one of the equations, but not introduced by solution.

The lawful cases determinable *directly* or *indirectly* by the Aristotelian Rules, are AI, EI, Fig. 1; AO, EI, OA, IE, Fig. 2; AI, AO, EI, EO, IA, IE, OA, OE, Fig. 3; IA, IE, Fig. 4.

Those not so determinable are OE, Fig. 1; EO, Fig. 4.

The cases in which no inference is possible, are AO, EO, IA, IE, OA, Fig. 1; AI, EO, IA, OE, Fig. 2; OA, OE, AI, EI, AO, Fig. 4.

Ex. 1. AI, Fig. 1, and, by mutation, IA, Fig. 4.

$$
\begin{array}{ll}
\text{All Ys are Xs,} & y\,(1 - x) = 0 \\
\text{Some Zs are Ys,} & vz\ \ = vy \\
\hline
& vz\,(1 - x) = 0 \\
\end{array}
$$

$\therefore$ Some Zs are Xs.

Ex. 2. AO, Fig. 2, and, by mutation, OA, Fig. 2.

$$
\begin{array}{lll}
\text{All Xs are Ys,} & x\,(1 - y) = 0, & x = xy \\
\text{Some Zs are not Ys,} & vz = v\,(1 - y), & vy = v\,(1 - z) \\
& & \overline{vx = vx\,(1 - z)} \\
& & vxz = 0 \\
\end{array}
$$

$\therefore$ Some Zs are not Xs.

The interpretation of vz as Some Zs, is implied, it will be observed, in the equation $vz = v\,(1 - y)$ considered as representing the proposition Some Zs are not Ys.

The cases not determinable by the Aristotelian Rules are OE, Fig. 1, and, by mutation, EO, Fig. 4.

$$
\begin{array}{ll}
\text{Some Ys are not Xs,} & vy = v\,(1 - x) \\
\text{No Zs are Ys,} & 0 = zy \\
\hline
& 0 = v\,(1 - x)\,z \\
\end{array}
$$

$\therefore$ Some not-Xs are not Zs.

The equation of the first premiss here permits us to interpret $v\,(1 - x)$, but it does not enable us to interpret vz.

Of cases in which no inference is possible, we take as examples—

AO, Fig. 1, and, by mutation, OA, Fig. 4,

All Ys are Xs,	$y(1-x)=0,$		$y(1-x)=0$
Some Zs are not Ys,	$vz=v(1-y)$	(a)	$v(1-z)=vy$

$$v(1-z)(1-x)=0 \quad (b)$$
$$0=0$$

since the auxiliary equation in this case is $v(1-z)=0$.

Practically it is not necessary to perform this reduction, but it is satisfactory to do so. The equation (a), it is seen, defines vz as Some Zs, but it does not define $v(1-z)$, so that we might stop at the result of elimination (b), and content ourselves with saying, that it is not interpretable into a relation between the classes X and Z.

Take as a second example AI, Fig. 2, and, by mutation, IA, Fig. 2,

All Xs are Ys,	$x(1-y)=0,$	$x=xy$
Some Zs are Ys,	$vz=vy,$	$vy=vz$

$$vx=vxz$$
$$v(1-z)x=0$$
$$0=0,$$

the auxiliary equation in this case being $v(1-z)=0$.

Indeed in every case in this class, in which no inference is possible, the result of elimination is reducible to the form $0=0$. Examples therefore need not be multiplied.

CLASS 4th.—When v enters into both equations.

No inference is possible in any case, but there exists a distinction among the unlawful cases which is peculiar to this class. The two divisions are,

1st. When the result of elimination is reducible by the auxiliary equations to the form $0=0$. The cases are II, OI,

Fig. 1; II, OO, Fig. 2; II, IO, OI, OO, Fig. 3; II, IO, Fig. 4.

2nd. When the result of elimination is not reducible by the auxiliary equations to the form 0 = 0.

The cases are IO, OO, Fig. 1; IO, OI, Fig. 2; OI, OO, Fig. 4.

Let us take as an example of the former case, II, Fig. 3.

$$\text{Some Xs are Ys,} \qquad vx = vy, \qquad vx = vy$$
$$\text{Some Zs are Ys,} \qquad v'z = v'y, \qquad \frac{v'y = v'z}{vv'x = vv'z}$$

Now the auxiliary equations $v(1 - x) = 0$, $v'(1 - z) = 0$,

$$\text{give} \quad vx = v, \quad v'z = v'.$$

Substituting we have

$$vv' = vv',$$
$$\therefore \ 0 = 0.$$

As an example of the latter case, let us take IO, Fig. 1,

$$\text{Some Ys are Xs,} \qquad vy = vx, \qquad vy = vx$$
$$\text{Some Zs are not Ys,} \quad v'z = v'(1 - y), \qquad \frac{v'(1 - z) = v'y}{vv'(1 - z) = vv'x}$$

Now the auxiliary equations being $v(1 - x) = 0$, $v'(1 - z) = 0$, the above reduces to $vv' = 0$. It is to this form that all similar cases are reducible. Its interpretation is, that the classes v and v' have no common member, as is indeed evident.

The above classification is purely founded on mathematical distinctions. We shall now inquire what is the logical division to which it corresponds.

The lawful cases of the first class comprehend all those in which, from two universal premises, a universal conclusion may be drawn. We see that they include the premises of *barbara* and *celarent* in the first figure, of *cesare* and *camestres* in the second, and of *bramantip* and *camenes* in the fourth.

The premises of *bramantip* are included, because they admit of an universal conclusion, although not in the same figure.

The lawful cases of the second class are those in which a particular conclusion only is deducible from two universal premises.

The lawful cases of the third class are those in which a conclusion is deducible from two premises, one of which is universal and the other particular.

The fourth class has no lawful cases.

Among the cases in which no inference of any kind is possible, we find six in the fourth class distinguishable from the others by the circumstance, that the result of elimination does not assume the form $0 = 0$.   The cases are

$$\begin{cases} \text{Some Ys are Xs,} \\ \text{Some Zs are not Ys,} \end{cases} \begin{cases} \text{Some Ys are not Xs,} \\ \text{Some Zs are not Ys,} \end{cases} \begin{cases} \text{Some Xs are Ys,} \\ \text{Some Zs are not Ys,} \end{cases}$$

and the three others which are obtained by mutation of premises.

It might be presumed that some logical peculiarity would be found to answer to the mathematical peculiarity which we have noticed, and in fact there exists a very remarkable one. If we examine each pair of premises in the above scheme, we shall find that there *is virtually* no middle term, i.e. *no medium of comparison*, in any of them. Thus, in the first example, the individuals spoken of in the first premiss are asserted to belong to the class Y, but those spoken of in the second premiss are *virtually* asserted to belong to the class not-Y: nor can we by any lawful transformation or conversion alter this state of things. The comparison will still be made with the class Y in one premiss, and with the class not-Y in the other.

Now in every case beside the above six, there will be found a middle term, either expressed or implied. I select two of the most difficult cases.

In AO, Fig. 1, viz.

> All Ys are Xs,
>
> Some Zs are not Ys,

we have, by *negative conversion* of the first premiss,

> All not-Xs are not-Ys,
>
> Some Zs are not Ys,

and the middle term is now seen to be not-Y.

Again, in EO, Fig. 1,

> No Ys are Xs,
>
> Some Zs are not Ys,

a proved conversion of the first premiss (see *Conversion of Propositions*), gives

> All Xs are not-Ys,
>
> Some Zs are not-Ys,

and the middle term, the true medium of comparison, is plainly not-Y, although as the not-Ys in the one premiss *may be* different from those in the other, no conclusion can be drawn.

The mathematical condition in question, therefore,—the irreducibility of the final equation to the form $0 = 0$,—adequately represents the logical condition of there being no middle term, or common medium of comparison, in the given premisses.

I am not aware that the distinction occasioned by the presence or absence of a middle term, in the strict sense here understood, has been noticed by logicians before. The distinction, though real and deserving attention, is indeed by no means an obvious one, and it would have been unnoticed in the present instance but for the peculiarity of its mathematical expression.

What appears to be novel in the above case is the proof of the existence of combinations of premisses in which there is absolutely no medium of comparison. When such a medium of comparison, or true middle term, does exist, the condition that its quantification in both premises together shall exceed its quantification as a single whole, has been ably and clearly shewn by Professor De Morgan to be necessary to lawful inference (*Cambridge Memoirs,* Vol. VIII. Part 3). And this is undoubtedly the true principle of the Syllogism, viewed from the standing-point of Arithmetic.

I have said that it would be possible to impose conditions of interpretation which should restrict the results of this calculus to the Aristotelian forms. Those conditions would be,

1st. That we should agree not to interpret the forms $v(1 - x)$, $v(1 - z)$.

2ndly. That we should agree to reject every interpretation in which the order of the terms thould violate the Aristotelian rule.

Or, instead of the second condition, it might be agreed that, the conclusion being determined, the order of the premises should, if necessary, be changed, so as to make the syllogism formal.

From the *general* character of the system it is indeed plain, that it may be made to represent any conceivable scheme of logic, by imposing the conditions proper to the case contemplated.

We have found it, in a certain class of cases, to be necessary to replace the two equations expressive of universal Propositions, by their solutions; and it may be proper to remark, that it would have been allowable in all instances to have done this,* so that every case of the Syllogism, without ex-

---

* It may be satisfactory to illustrate this statement by an example. In *Barbara*, we should have

$$\begin{aligned}
\text{All Ys are Xs,} && y &= vx \\
\text{All Zs are Ys,} && \underline{z} &\underline{= v'y} \\
&& z &= vv'x
\end{aligned}$$

$$\therefore \text{ All Zs are Xs.}$$

ception, might have been treated by equations comprised in the general forms

$$y = vx, \qquad \text{or} \quad y - vx = 0 \; \dots \; \text{A},$$
$$y = v\,(1 - x), \qquad \text{or} \; y + vx - v = 0 \; \dots \; \text{E},$$
$$vy = vx, \qquad\qquad vy - vx = 0 \; \dots \; \text{I},$$
$$vy = v\,(1 - x), \qquad\quad vy + vx - v = 0 \; \dots \; \text{O}.$$

Or, we may multiply the resulting equation by $1 - x$, which gives

$$z\,(1 - x) = 0,$$

whence the same conclusion, All Zs are Xs.

Some additional examples of the application of the system of equations in the text to the demonstration of general theorems, may not be inappropriate.

Let y be the term to be eliminated, and let x stand indifferently for either of the other symbols, then each of the equations of the premises of any given syllogism may be put in the form

$$ay + bx = 0, \quad (a)$$

if the premiss is affirmative, and in the form

$$ay + b\,(1 - x) = 0, \quad (\beta)$$

if it is negative, a and b being either constant, or of the form $\pm v$. To prove this in detail, let us examine each kind of proposition, making y successively subject and predicate.

A,	All Ys are Xs,	$y - vx = 0,$	$(\gamma),$
	All Xs are Ys,	$x - vy = 0,$	$(\delta),$
E,	No Ys are Xs,	$xy = 0,$	
	No Xs are Ys,	$y - v\,(1 - x) = 0,$	$(\varepsilon),$
I,	Some Xs are Ys,		
	Some Ys are Xs,	$vx - vy = 0,$	$(\zeta),$
O,	Some Ys are not Xs,	$vy - v\,(1 - x) = 0,$	$(\eta),$
	Some Xs are not Ys,	$vx = v\,(1 - y),$	
		$\therefore vy - v\,(1 - x) = 0,$	$(\theta).$

The affirmative equations (γ), (δ) and (ζ), belong to (a), and the negative equations (ε), (η) and (θ), to (β). It is seen that the two last negative equations are alike, but there is a difference of interpretation. In the former

$$v\,(1 - x) = \text{Some not-Xs},$$

in the latter,

$$v\,(1 - x) = 0.$$

The utility of the two general forms of reference, (a) and (β), will appear from the following application.

1st. *A conclusion drawn from two affirmative propositions* is itself affirmative.

By (a) we have for the given propositions,

$$ay + bx = 0,$$
$$a'y + b'z = 0,$$

Perhaps the system we have actually employed is better, as distinguishing the cases in which v only *may* be employed, and eliminating

$$ab'z - a'bx = 0,$$

which is of the form (α). Hence, if there is a conclusion, it is affirmative.

2nd. *A conclusion drawn from an affirmative and a negative proposition is negative.*

By (α) and (β), we have for the given propositions

$$ay + bx = 0,$$
$$a'y + b'(1 - z) = 0,$$
$$\therefore a'bx - ab'(1 - z) = 0,$$

which is of the form (β). Hence the conclusion, if there is one, is negative.

3rd. *A conclusion drawn from two negative premises will involve a negation, (not-X, not-Z) in both subject and predicate, and will therefore be inadmissible in the Aristotelian system, though just in itself.*

For the premises being

$$ay + b(1 - x) = 0,$$
$$a'y + b'(1 - z) = 0,$$

the conclusion will be

$$ab'(1 - z) - a'b(1 - x) = 0,$$

which is only interpretable into a proposition that has a negation in each term.

4th. *Taking into account those syllogisms only, in which the conclusion is the most general, that can be deduced from the premises,—if, in an Aristotelian syllogism, the minor premises be changed in quality (from affirmative to negative or from negative to affirmative), whether it be changed in quantity or not, no conclusion will be deducible in the same figure.*

An Aristotelian proposition does not admit a term of the form not-Z in the subject,—Now on changing the quantity of the minor proposition of a syllogism, we transfer it from the general form

$$ay + bz = 0,$$

to the general form

$$a'y + b'(1 - z) = 0,$$

see (α) *and* (β), or *vice versâ*. And therefore, in the equation of the conclusion, there will be a change from z to $1 - z$, or *vice versâ*. But this is equivalent to the change of Z into not-Z, or not-Z into Z. Now the subject of the original conclusion must have involved a Z and not a not-Z, therefore the subject of the new conclusion will involve a not-Z, and the conclusion will not be admissible in the Aristotelian forms, except by conversion, which would render necessary a change of Figure.

Now the conclusions of this calculus are always the most general that can be drawn, and therefore the above demonstration must not be supposed to extend to a syllogism, in which a particular conclusion is deduced, when a universal one is possible. This is the case with *bramantip* only, among the Aristotelian forms, and therefore the transformation of *bramantip* into *camenes*, and *vice versâ*, is the case of restriction contemplated in the preliminary statement of the theorem.

from those in which it *must*. But for the demonstration of certain general properties of the Syllogism, the above system is, from its simplicity, and from the mutual analogy of its forms, very convenient. We shall apply it to the following theorem.*

Given the three propositions of a Syllogism, prove that there is but one order in which they can be legitimately arranged, and determine that order.

All the forms above given for the expression of propositions, are particular cases of the general form,

$$a + bx + cy = 0.$$

5th. *If for the minor premiss of an Aristotelian syllogism, we substitute its contradictory, no conclusion is deducible in the same figure.*

It is here only necessary to examine the case of *bramantip*, all the others being determined by the last proposition.

On changing the minor of *bramantip* to its contradictory, we have AO, Fig. 4, and this admits of no legitimate inference.

Hence the theorem is true without exception. Many other general theorems may in like manner be proved.

* This elegant theorem was communicated by the Rev. Charles Graves, Fellow and Professor of Mathematics in Trinity College, Dublin, to whom the Author desires further to record his grateful acknowledgments for a very judicious examination of the former portion of this work, and for some new applications of the method. The following example of Reduction *ad impossibile* is among the number :

Reducend Mood,	All Xs are Ys,	$1 - y = v'(1 - x)$
Baroko	Some Zs are not Ys	$vz = v(1 - y)$
	Some Zs are not Xs	$vz = vv'(1 - x)$
Reduct Mood,	All Xs are Ys	$1 - y = v'(1 - x)$
Barbara	All Zs are Xs	$z(1 - x) = 0$
	All Zs are Ys	$z(1 - y) = 0.$

The conclusion of the reduct mood is seen to be the contradictory of the suppressed minor premiss. Whence, &c. It may just be remarked that the mathematical test of contradictory propositions is, that on eliminating one elective symbol between their equations, the other elective symbol vanishes. The *ostensive* reduction of *Baroko* and *Bokardo* involves no difficulty.

Professor Graves suggests the employment of the equation $x = vy$ for the primary expression of the Proposition All Xs are Ys, and remarks, that on multiplying both members by $1 - y$, we obtain $x(1 - y) = 0$, the equation from which we set out in the text, and of which the previous one is a solution.

Assume then for the premises of the given syllogism, the equations

$$a + bx + cy = 0, \quad (18),$$

$$a' + b'z + c'y = 0, \quad (19),$$

then, eliminating y, we shall have for the conclusion

$$ac' - a'c + bc'x - b'cz = 0, \quad (20).$$

Now taking this as one of our premises, and either of the original equations, suppose (18), as the other, if by elimination of a common term x, between them, we can obtain a result equivalent to the remaining premiss (19), it will appear that there are more than one order in which the Propositions may be lawfully written; but if otherwise, one arrangement only is lawful.

Effecting then the elimination, we have

$$bc\,(a' + b'z + c'y) = 0, \quad (21),$$

which is equivalent to (19) multiplied by a factor bc. Now on examining the value of this factor in the equations A, E, I, O, we find it in each case to be v or $-v$. But it is evident, that if an equation expressing a given Proposition be multiplied by an extraneous factor, derived from another equation, its interpretation will either be limited or rendered impossible. Thus there will either be no result at all, or the result will be a *limitation* of the remaining Proposition.

If, however, one of the original equations were

$$x = y, \quad \text{or} \quad x - y = 0,$$

the factor bc would be -1, and would *not* limit the interpretation of the other premiss. Hence if the first member of a syllogism should be understood to represent the double proposition All Xs are Ys, and All Ys are Xs, it would be indifferent in what order the remaining Propositions were written.

A more general form of the above investigation would be, to express the premises by the equations

$$a + bx + cy + dxy = 0, \quad (22),$$
$$a' + b'z + c'y + d'zy = 0, \quad (23).$$

After the double elimination of y and x we should find

$$(bc - ad)\,(a + b'z + c'y + d'zy) = 0\,;$$

and it would be seen that the factor $bc - ad$ must in every case either vanish or express a limitation of meaning.

The determination of the order of the Propositions is sufficiently obvious.

OF HYPOTHETICALS.

————————

A hypothetical Proposition is defined to be *two or more categoricals united by a copula* (or conjunction), and the different kinds of hypothetical Propositions are named from their respective conjunctions, viz. conditional (if), disjunctive (either, or), &c.

In conditionals, that categorical Proposition from which the other results is called the *antecedent*, that which results from it the *consequent*.

Of the conditional syllogism there are two, and only two formulæ.

1st. The constructive,

If A is B, then C is D,

But A is B, therefore C is D.

2nd. The Destructive,

If A is B, then C is D,

But C is not D, therefore A is not B.

A dilemma is a complex conditional syllogism, with several antecedents in the major, and a disjunctive minor.

IF we examine either of the forms of conditional syllogism above given, we shall see that the validity of the argument does not depend upon any considerations which have reference to the terms A, B, C, D, considered as the representatives of individuals or of classes. We may, in fact, represent the Propositions A is B, C is D, by the arbitrary symbols X and Y respectively, and express our syllogisms in such forms as the following:

If X is true, then Y is true,

But X is true, therefore Y is true.

Thus, what we have to consider is not objects and classes of objects, but the truths of Propositions, namely, of those elementary Propositions which are embodied in the terms of our hypothetical premises.

To the symbols X, Y, Z, representative of Propositions, we may appropriate the elective symbols x, y, z, in the following sense.

The hypothetical Universe, 1, shall comprehend all conceivable cases and conjunctures of circumstances.

The elective symbol x attached to any subject expressive of such cases shall select those cases in which the Proposition X is true, and similarly for Y and Z.

If we confine ourselves to the contemplation of a given proposition X, and hold in abeyance every other consideration, then two cases only are conceivable, viz. first that the given Proposition is true, and secondly that it is false.* As these cases together make up the Universe of the Proposition, and as the former is determined by the elective symbol x, the latter is determined by the symbol $1 - x$.

But if other considerations are admitted, each of these cases will be resolvable into others, individually less extensive, the

---

* It was upon the obvious principle that a Proposition is either true or false, that the Stoics, applying it to assertions respecting future events, endeavoured to establish the doctrine of Fate. It has been replied to their argument, that it involves "an abuse of the word *true*, the precise meaning of which is id quod res *est*. An assertion respecting the future is neither true nor false."—*Copleston on Necessity and Predestination*, p. 36. Were the Stoic axiom, however, presented under the form, It is either certain that a given event will take place, or certain that it will not; the above reply would fail to meet the difficulty. The proper answer would be, that no merely verbal definition can settle the question, what is the actual course and constitution of Nature. When we affirm that it is either certain that an event will take place, or certain that it will not take place, we tacitly assume that the order of events is necessary, that the Future is but an evolution of the Present; so that the state of things which is, completely determines that which shall be. But this (at least as respects the conduct of moral agents) is the very question at issue. Exhibited under its proper form, the Stoic reasoning does not involve an abuse of terms, but a *petitio principii*.

It should be added, that enlightened advocates of the doctrine of Necessity in the present day, viewing the end as appointed only in and through the means, justly repudiate those practical ill consequences which are the reproach of Fatalism.

number of which will depend upon the number of foreign considerations admitted. Thus if we associate the Propositions X and Y, the total number of conceivable cases will be found as exhibited in the following scheme.

Cases.		Elective expressions.
1st	X true, Y true	xy
2nd	X true, Y false............	$x(1-y)$
3rd	X false, Y true............	$(1-x)y$
4th	X false, Y false............	$(1-x)(1-y)$ (24).

If we add the elective expressions for the two first of the above cases the sum is x, which is the elective symbol appropriate to the more general case of X being true independently of any consideration of Y; and if we add the elective expressions in the two last cases together, the result is $1-x$, which is the elective expression appropriate to the more general case of X being false.

Thus the extent of the hypothetical Universe does not at all depend upon the number of circumstances which are taken into account. And it is to be noted that however few or many those circumstances may be, the sum of the elective expressions representing every conceivable case will be unity. Thus let us consider the three Propositions, X, It rains, Y, It hails, Z, It freezes. The possible cases are the following:

Cases.		Elective expressions.
1st	It rains, hails, and freezes,	xyz
2nd	It rains and hails, but does not freeze	$xy(1-z)$
3rd	It rains and freezes, but does not hail	$xz(1-y)$
4th	It freezes and hails, but does not rain	$yz(1-x)$
5th	It rains, but neither hails nor freezes	$x(1-y)(1-z)$
6th	It hails, but neither rains nor freezes	$y(1-x)(1-z)$
7th	It freezes, but neither hails nor rains	$z(1-x)(1-y)$
8th	It neither rains, hails, nor freezes	$(1-x)(1-y)(1-z)$

$$1 = \text{sum}$$

Expression of Hypothetical Propositions.

To express that a given Proposition X is true.

The symbol $1 - x$ selects those cases in which the Proposition X is false. But if the Proposition is true, there are no such cases in its hypothetical Universe, therefore

$$1 - x = 0,$$

or
$$x = 1, \quad (25).$$

To express that a given Proposition X is false.

The elective symbol x selects all those cases in which the Proposition is true, and therefore if the Proposition is false,

$$x = 0, \quad (26).$$

And in every case, having determined the elective expression appropriate to a given Proposition, we assert the truth of that Proposition by equating the elective expression to unity, and its falsehood by equating the same expression to 0.

To express that two Propositions, X and Y, are simultaneously true.

The elective symbol appropriate to this case is xy, therefore the equation sought is

$$xy = 1, \quad (27).$$

To express that two Propositions, X and Y, are simultaneously false.

The condition will obviously be

$$(1 - x)(1 - y) = 1,$$

or
$$x + y - xy = 0, \quad (28).$$

To express that either the Proposition X is true, or the Proposition Y is true.

To assert that either one or the other of two Propositions is true, is to assert that it is not true, that they are both false. Now the elective expression appropriate to their both being false is $(1 - x)(1 - y)$, therefore the equation required is

$$(1 - x)(1 - y) = 0,$$

or
$$x + y - xy = 1, \quad (29),$$

E

And, by indirect considerations of this kind, may every disjunctive Proposition, however numerous its members, be expressed. But the following general Rule will usually be preferable.

RULE. *Consider what are those distinct and mutually exclusive cases of which it is implied in the statement of the given Proposition, that some one of them is true, and equate the sum of their elective expressions to unity. This will give the equation of the given Proposition.*

For the sum of the elective expressions for all distinct conceivable cases will be unity. Now all these cases being mutually exclusive, and it being asserted in the given Proposition that some one case out of a given set of them is true, it follows that all which are not included in that set are false, and that their elective expressions are severally equal to 0. Hence the sum of the elective expressions for the remaining cases, viz. those included in the given set, will be unity. Some one of those cases will therefore be true, and as they are mutually exclusive, it is impossible that more than one should be true. Whence the Rule in question.

And in the application of this Rule it is to be observed, that if the cases contemplated in the given disjunctive Proposition are not mutually exclusive, they must be resolved into an equivalent series of cases which are mutually exclusive.

Thus, if we take the Proposition of the preceding example, viz. Either X is true, or Y is true, and assume that the two members of this Proposition are not exclusive, insomuch that in the enumeration of possible cases, we must reckon that of the Propositions X and Y being both true, then the mutually exclusive cases which fill up the Universe of the Proposition, with their elective expressions, are

1st, X true and Y false,	$x(1-y)$,
2nd, Y true and X false,	$y(1-x)$,
3rd, X true and Y true,	xy,

and the sum of these elective expressions equated to unity gives

$$x + y - xy = 1. \quad (30),$$

as before. But if we suppose the members of the disjunctive Proposition to be exclusive, then the only cases to be considered are

1st,	X true, Y false,	$x\,(1 - y)$,
2nd,	Y true, X false,	$y\,(1 - x)$,

and the sum of these elective expressions equated to 0, gives

$$x - 2xy + y = 1, \quad (31).$$

The subjoined examples will further illustrate this method.

To express the Proposition, Either X is not true, or Y is not true, the numbers being exclusive.

The mutually exclusive cases are

1st,	X not true, Y true,	$y\,(1 - x)$,
2nd,	Y not true, X true,	$x\,(1 - y)$,

and the sum of these equated to unity gives

$$x - 2xy + y = 1, \quad (32),$$

which is the same as (31), and in fact the Propositions which they represent are equivalent.

To express the Proposition, Either X is not true, or Y is not true, the members not being exclusive.

To the cases contemplated in the last Example, we must add the following, viz.

X not true, Y not true, $(1 - x)\,(1 - y)$.

The sum of the elective expressions gives

$$x\,(1 - y) + y\,(1 - x) + (1 - x)\,(1 - y) = 1,$$
$$\text{or} \quad xy = 0, \quad (33).$$

To express the disjunctive Proposition, Either X is true, or Y is true, or Z is true, the members being exclusive.

Here the mutually exclusive cases are

1st,	X true, Y false, Z false,	$x(1-y)(1-z)$,
2nd,	Y true, Z false, X false,	$y(1-z)(1-x)$,
3rd,	Z true, X false, Y false,	$z(1-x)(1-y)$,

and the sum of the elective expressions equated to 1, gives, upon reduction,

$$x + y + z - 2(xy + yz + zx) + 3xyz = 1, \quad (34).$$

The expression of the same Proposition, when the members are in no sense exclusive, will be

$$(1-x)(1-y)(1-z) = 0, \quad (35).$$

And it is easy to see that our method will apply to the expression of any similar Proposition, whose members are subject to any specified amount and character of exclusion.

To express the conditional Proposition, If X is true, Y is true.

Here it is implied that all the cases of X being true, are cases of Y being true. The former cases being determined by the elective symbol x, and the latter by y, we have, in virtue of (4),

$$x(1-y) = 0, \quad (36).$$

To express the conditional Proposition, If X be true, Y is not true.

The equation is obviously

$$xy = 0, \quad (37);$$

this is equivalent to (33), and in fact the disjunctive Proposition, Either X is not true, or Y is not true, and the conditional Proposition, If X is true, Y is not true, are equivalent.

To express that If X is not true, Y is not true.

In (36) write $1-x$ for x, and $1-y$ for y, we have

$$(1-x)y = 0.$$

The results which we have obtained admit of verification in many different ways. Let it suffice to take for more particular examination the equation

$$x - 2xy + y = 1, \quad (38),$$

which expresses the conditional Proposition, Either X is true, or Y is true, the members being in this case exclusive.

First, let the Proposition X be true, then $x = 1$, and substituting, we have

$$1 - 2y + y = 1, \quad \therefore - y = 0, \quad \text{or} \quad y = 0,$$

which implies that Y is not true.

Secondly, let X be not true, then $x = 0$, and the equation gives

$$y = 1, \quad (39),$$

which implies that Y is true. In like manner we may proceed with the assumptions that Y is true, or that Y is false.

Again, in virtue of the property $x^2 = x$, $y^2 = y$, we may write the equation in the form

$$x^2 - 2xy + y^2 = 1,$$

and extracting the square root, we have

$$x - y = \pm 1, \quad (40),$$

and this represents the actual case; for, as when X is true or false, Y is respectively false or true, we have

$$x = 1 \text{ or } 0,$$

$$y = 0 \text{ or } 1,$$

$$\therefore x - y = 1 \text{ or } - 1.$$

There will be no difficulty in the analysis of other cases.

Examples of Hypothetical Syllogism.

The treatment of every form of hypothetical Syllogism will consist in forming the equations of the premises, and eliminating the symbol or symbols which are found in more than one of them. The result will express the conclusion.

1st. Disjunctive Syllogism.

Either X is true, or Y is true (exclusive), $x + y - 2\,xy = 1$
But X is true, $x = 1$
Therefore Y is not true, $\therefore \; y = 0$

Either X is true, or Y is true (not exclusive), $x + y - xy = 1$
But X is not true, $x = 0$
Therefore Y is true, $\therefore \; y = 1$

2nd. Constructive Conditional Syllogism.

If X is true, Y is true, $x\,(1 - y) = 0$
But X is true, $x = 1$
Therefore Y is true, $\therefore \; 1 - y = 0$ or $y = 1$.

3rd. Destructive Conditional Syllogism.

If X is true, Y is true, $x\,(1 - y) = 0$
But Y is not true, $y = 0$
Therefore X is not true, $\therefore \; x = 0$

4th. Simple Constructive Dilemma, the minor premiss exclusive.

If X is true, Y is true, $x\,(1 - y) = 0$,   (41),
If Z is true, Y is true, $z\,(1 - y) = 0$,   (42),
But Either X is true, or Z is true, $x + z - 2xz = 1$,   (43).

From the equations (41), (42), (43), we have to eliminate x and z. In whatever way we effect this, the result is

$$y = 1;$$

whence it appears that the Proposition Y is true.

5th. Complex Constructive Dilemma, the minor premiss not exclusive.

If X is true, Y is true, $x\,(1 - y) = 0$,
If W is true, Z is true, $w\,(1 - z) = 0$,
Either X is true, or W is true, $x + w - xw = 1$.

From these equations, eliminating x, we have

$$y + z - yz = 1,$$

which expresses the Conclusion, Either Y is true, or Z is true, the members being non-exclusive.

6th. Complex Destructive Dilemma, the minor premiss exclusive.

If X is true, Y is true,	$x(1-y)=0$
If W is true, Z is true,	$w(1-z)=0$
Either Y is not true, or Z is not true,	$y+z-2yz=1.$

From these equations we must eliminate y and z. The result is $$xw = 0,$$

which expresses the Conclusion, Either X is not true, or Y is not true, the members *not being exclusive.*

7th. Complex Destructive Dilemma, the minor premiss not exclusive.

If X is true, Y is true,	$x(1-y)=0$
If W is true, Z is true,	$w(1-z)=0$
Either Y is not true, or Z is not true,	$yz=0.$

On elimination of y and z, we have

$$xw = 0,$$

which indicates the same Conclusion as the previous example.

It appears from these and similar cases, that whether the members of the minor premiss of a Dilemma are exclusive or not, the members of the (disjunctive) Conclusion are never exclusive. This fact has perhaps escaped the notice of logicians.

The above are the principal forms of hypothetical Syllogism which logicians have recognised. It would be easy, however, to extend the list, especially by the blending of the disjunctive and the conditional character in the same Proposition, of which the following is an example.

If X is true, then either Y is true, or Z is true,
$$x(1-y-z+yz)=0$$
But Y is not true, $y=0$
Therefore If X is true, Z is true, $\therefore x(1-z)=0.$

That which logicians term a *Causal* Proposition is properly a conditional Syllogism, the major premiss of which is suppressed.

The assertion that the Proposition X is true, *because* the Proposition Y is true, is equivalent to the assertion,

> The Proposition Y is true,
> *Therefore* the Proposition X is true;

and these are the minor premiss and conclusion of the conditional Syllogism,

> If Y is true, X is true,
> But Y is true,
> Therefore X is true.

And thus causal Propositions are seen to be included in the applications of our general method.

Note, that there is a family of disjunctive and conditional Propositions, which do not, of right, belong to the class considered in this Chapter. Such are those in which the force of the disjunctive or conditional particle is expended upon the predicate of the Proposition, as if, speaking of the inhabitants of a particular island, we should say, that they are all *either Europeans or Asiatics;* meaning, that it is true of each individual, that he is either a European or an Asiatic. If we appropriate the elective symbol x to the inhabitants, y to Europeans, and z to Asiatics, then the equation of the above Proposition is

$$x = xy + xz, \quad \text{or} \quad x(1 - y - z) = 0, \quad (a);$$

to which we might add the condition $yz = 0$, since no Europeans are Asiatics. The nature of the symbols x, y, z, indicates that the Proposition belongs to those which we have before designated as *Categorical*. Very different from the above is the Proposition, Either all the inhabitants are Europeans, or they are all Asiatics. Here the disjunctive particle separates Propositions. The case is that contemplated in (31) of the present Chapter; and the symbols by which it is expressed, although subject to the same laws as those of (*a*), have a totally different interpretation.*

The distinction is real and important. Every Proposition which language can express may be represented by elective symbols, and the laws of combination of those symbols are in all cases the same; but in one class of instances the symbols have reference to collections of objects, in the other, to the truths of constituent Propositions.

---

* Some writers, among whom is Dr. Latham (*First Outlines*), regard it as the exclusive office of a conjunction to connect *Propositions*, not *words*. In this view I am not able to agree. The Proposition, Every animal is *either* rational *or* irrational, cannot be resolved into, *Either* every animal is rational, *or* every animal is irrational. The former belongs to pure categoricals, the latter to hypotheticals. In *singular* Propositions, such conversions would seem to be allowable. This animal is *either* rational *or* irrational, is equivalent to, *Either* this animal is rational, *or* it is irrational. This peculiarity of *singular* Propositions would almost justify our ranking them, though truly universals, in a separate class, as Ramus and his followers did.

PROPERTIES OF ELECTIVE FUNCTIONS.

SINCE elective symbols combine according to the laws of quantity, we may, by Maclaurin's theorem, expand a given function $\phi(x)$, in ascending powers of x, known cases of failure excepted. Thus we have

$$\phi(x) = \phi(0) + \phi'(0)\, x + \frac{\phi''(0)}{1.2}\, x^2 + \&c., \quad (44).$$

Now $x^2 = x,\ x^3 = x,\ \&c.,$ whence

$$\phi(x) = \phi(0) + x \left\{ \phi'(0) + \frac{\phi''(0)}{1.2} + \&c. \right\}, \quad (45).$$

Now if in (44) we make $x = 1$, we have

$$\phi(1) = \phi(0) + \phi'(0) + \frac{\phi''(0)}{1.2} + \&c.,$$

whence

$$\phi'(0) + \frac{\phi''(0)}{1.2} + \frac{\phi'''(0)}{1.2.3} + \&c. = \phi(1) - \phi(0).$$

Substitute this value for the coefficient of x in the second member of (45), and we have*

$$\phi(x) = \phi(0) + \left\{ \phi(1) - \phi(0) \right\} x, \quad (46),$$

* Although this and the following theorems have only been proved for those forms of functions which are expansible by Maclaurin's theorem, they may be regarded as true for all forms whatever; this will appear from the applications. The reason seems to be that, as it is only through the one form of expansion that elective functions become interpretable, no conflicting interpretation is possible.

The development of $\phi(x)$ may also be determined thus. By the known formula for expansion in factorials,

$$\phi(x) = \phi(0) + \Delta\phi(0)\, x + \frac{\Delta^2 \phi(0)}{1.2}\, x\,(x-1) + \&c.$$

which we shall also employ under the form

$$\phi(x) = \phi(1) x + \phi(0)(1-x), \quad (47).$$

Every function of x, in which integer powers of that symbol are alone involved, is by this theorem reducible to the first order. The quantities $\phi(0)$, $\phi(1)$, we shall call the moduli of the function $\phi(x)$. They are of great importance in the theory of elective functions, as will appear from the succeeding Propositions.

PROP. 1. Any two functions $\phi(x)$, $\psi(x)$, are equivalent, whose corresponding moduli are equal.

This is a plain consequence of the last Proposition. For since

$$\phi(x) = \phi(0) + \{\phi(1) - \phi(0)\} x,$$
$$\psi(x) = \psi(0) + \{\psi(1) - \psi(0)\} x,$$

it is evident that if $\phi(0) = \psi(0)$, $\phi(1) = \psi(1)$, the two expansions will be equivalent, and therefore the functions which they represent will be equivalent also.

The converse of this Proposition is equally true, viz.

If two functions are equivalent, their corresponding moduli are equal.

Among the most important applications of the above theorem, we may notice the following.

Suppose it required to determine for what forms of the function $\phi(x)$, the following equation is satisfied, viz.

$$\{\phi(x)\}^n = \phi(x).$$

Now x being an elective symbol, $x(x-1) = 0$, so that all the terms after the second, vanish. Also $\Delta \phi(0) = \phi(1) - \phi(0)$, whence

$$\phi\{x = \phi(0)\} + \{\phi(1) - \phi(0)\}x.$$

The mathematician may be interested in the remark, that this is not the only case in which an expansion stops at the second term. The expansions of the compound operative functions $\phi\left(\dfrac{d}{dx} + x^{-1}\right)$ and $\phi\left\{x + \left(\dfrac{d}{dx}\right)^{-1}\right\}$ are,

respectively,

$$\phi\left(\frac{d}{dx}\right) + \phi'\left(\frac{d}{dx}\right) x^{-1},$$

and

$$\phi(x) + \phi'(x)\left(\frac{d}{dx}\right)^{-1}.$$

See *Cambridge Mathematical Journal*, Vol. IV. p. 219.

Here we at once obtain for the expression of the conditions in question,

$$\{\phi\,(0)\}^n = \phi\,(0). \quad \{\phi\,(1)\}^n = \phi\,(1), \quad (48).$$

Again, suppose it required to determine the conditions under which the following equation is satisfied, viz.

$$\phi\,(x)\,\psi\,(x) = \chi\,(x),$$

The general theorem at once gives

$$\phi\,(0)\,\psi\,(0) = \chi\,(0). \quad \phi\,(1)\,\psi\,(1) = \chi\,(1), \quad (49).$$

This result may also be proved by substituting for $\phi\,(x)$, $\psi\,(x)$, $\chi\,(x)$, their expanded forms, and equating the coefficients of the resulting equation properly reduced.

All the above theorems may be extended to functions of more than one symbol. For, as different elective symbols combine with each other according to the same laws as symbols of quantity, we can first expand a given function with reference to any particular symbol which it contains, and then expand the result with reference to any other symbol, and so on in succession, the order of the expansions being quite indifferent.

Thus the given function being $\phi\,(xy)$ we have

$$\phi\,(xy) = \phi\,(x0) + \{\phi\,(x1) - \phi\,(x0)\}\,y,$$

and expanding the coefficients with reference to x, and reducing

$$\phi\,(xy) = \phi\,(00) + \{\phi\,(10) - \phi\,(00)\}\,x + \{\phi\,(01) - \phi\,(00)\}y$$
$$+ \{\phi\,(11) - \phi\,(10) - \phi\,(01) + \phi\,(00)\}\,xy, \quad (50),$$

to which we may give the elegant symmetrical form

$$\phi\,(xy) = \phi\,(00)\,(1 - x)\,(1 - y) + \phi\,(01)\,y\,(1 - x)$$
$$+ \phi\,(10)\,x\,(1 - y) + \phi\,(11)\,xy, \quad (51),$$

wherein we shall, in accordance with the language already employed, designate $\phi\,(00)$, $\phi\,(01)$, $\phi\,(10)$, $\phi\,(11)$, as the moduli of the function $\phi\,(xy)$.

By inspection of the above general form, it will appear that any functions of two variables are equivalent, whose corresponding moduli are all equal.

Thus the conditions upon which depends the satisfaction of the equation,

$$\{\phi\,(xy)\}^n = \phi\,(xy)$$

are seen to be

$$\{\phi\,(00)\}^n = \phi\,(00), \qquad \{\phi\,(01)\}^n = \phi\,(01), \qquad (52).$$
$$\{\phi\,(10)\}^n = \phi\,(10), \qquad \{\phi\,(11)\}^n = \phi\,(11),$$

And the conditions upon which depends the satisfaction of the equation

$$\phi\,(xy)\,\psi\,(xy) = \chi\,(xy),$$

are

$$\phi\,(00)\,\psi\,(00) = \chi\,(00), \qquad \phi\,(01)\,\psi\,(01) = \chi\,(01), \qquad (53).$$
$$\phi\,(10)\,\psi\,(10) = \chi\,(10), \qquad \phi\,(11)\,\psi\,(11) = \chi\,(11),$$

It is very easy to assign by induction from (47) and (51), the general form of an expanded elective function. It is evident that if the number of elective symbols is m, the number of the moduli will be 2^m, and that their separate values will be obtained by interchanging in every possible way the values 1 and 0 in the places of the elective symbols of the given function. The several terms of the expansion of which the moduli serve as coefficients, will then be formed by writing for each 1 that recurs under the functional sign, the elective symbol x, &c., which it represents, and for each 0 the corresponding $1 - x$, &c., and regarding these as factors, the product of which, multiplied by the modulus from which they are obtained, constitutes a term of the expansion.

Thus, if we represent the moduli of any elective function $\phi\,(xy\ldots)$ by $a_1, a_2, \ldots a_r$, the function itself, when expanded and arranged with reference to the moduli, will assume the form

$$\phi\,(xy) = a_1 t_1 + a_2 t_2 \ldots + a_r t_r, \quad (54),$$

in which $t_1 t_2 \ldots t_r$ are functions of $x, y \ldots$, resolved into factors of the forms $x, y, \ldots 1 - x, 1 - y, \ldots$ &c. These functions satisfy individually the index relations

$$t_1^n = t_1, \quad t_2^n = t_2, \quad \&c. \quad (55),$$

and the further relations,

$$t_1 t_2 = 0 \ldots t_1 t_2 = 0, \quad \&c. \quad (56),$$

the product of any two of them vanishing. This will at once be inferred from inspection of the particular forms (47) and (51). Thus in the latter we have for the values of t_1, t_2, &c., the forms

$$xy, \quad x(1-y), \quad (1-x)y, \quad (1-x)(1-y);$$

and it is evident that these satisfy the index relation, and that their products all vanish. We shall designate $t_1 t_2$.. as the constituent functions of $\phi(xy)$, and we shall define the peculiarity of the vanishing of the binary products, by saying that those functions are *exclusive*. And indeed the classes which they represent are mutually exclusive.

The sum of all the constituents of an expanded function is unity. An elegant proof of this Proposition will be obtained by expanding 1 as a function of any proposed elective symbols. Thus if in (51) we assume $\phi(xy) = 1$, we have $\phi(11) = 1$,

$$\phi(10) = 1, \quad \phi(01) = 1, \quad \phi(00) = 1, \quad \text{and (51) gives}$$

$$1 = xy + x(1-y) + (1-x)y + (1-x)(1-y), \quad (57).$$

It is obvious indeed, that however numerous the symbols involved, all the moduli of unity are unity, whence the sum of the constituents is unity.

We are now prepared to enter upon the question of the general interpretation of elective equations. For this purpose we shall find the following Propositions of the greatest service.

PROP. 2. If the first member of the general equation $\phi(xy...) = 0$, be expanded in a series of terms, each of which is of the form at, a being a modulus of the given function, then for every numerical modulus a which does not vanish, we shall have the equation
$$at = 0,$$
and the combined interpretations of these several equations will express the full significance of the original equation.

For, representing the equation under the form

$$a_1 t_1 + a_2 t_2 .. + a_r t_r = 0, \quad (58).$$

Multiplying by t_1, we have, by (56),

$$a_1 t_1 = 0, \quad (59),$$

whence if a_1 is a numerical constant which does not vanish,
$$t_1 = 0,$$
and similarly for all the moduli which do not vanish. And inasmuch as from these constituent equations we can form the given equation, their interpretations will together express its entire significance.

Thus if the given equation were
$$x - y = 0, \quad \text{Xs and Ys are identical,} \quad (60),$$
we should have $\phi(11) = 0$, $\phi(10) = 1$, $\phi(01) = -1$, $\phi(00) = 0$, so that the expansion (51) would assume the form
$$x(1 - y) - y(1 - x) = 0,$$
whence, by the above theorem,
$$x(1 - y) = 0, \qquad \text{All Xs are Ys,}$$
$$y(1 - x) = 0, \qquad \text{All Ys are Xs,}$$
results which are together equivalent to (60).

It may happen that the simultaneous satisfaction of equations thus deduced, may require that one or more of the elective symbols should vanish. This would only imply the nonexistence of a class: it may even happen that it may lead to a final result of the form
$$1 = 0,$$
which would indicate the nonexistence of the logical Universe. Such cases will only arise when we attempt to unite contradictory Propositions in a single equation. The manner in which the difficulty seems to be evaded in the result is characteristic.

It appears from this Proposition, that the differences in the interpretation of elective functions depend solely upon the number and position of the vanishing moduli. No change in the value of a modulus, but one which causes it to vanish, produces any change in the interpretation of the equation in which it is found. If among the infinite number of different values which we are thus permitted to give to the moduli which do not vanish in a proposed equation, any one value should be preferred, it is unity, for when the moduli of a function are all either 0 or 1, the function itself satisfies the condition

$$\{\phi\,(xy\,.\,.)\}^n = \phi\,(xy\ldots),$$

and this at once introduces symmetry into our Calculus, and provides us with fixed standards for reference.

PROP. 3. If $w = \phi\,(xy\,.\,.)$, w, x, y, .. being elective symbols, and if the second member be completely expanded and arranged in a series of terms of the form at, we shall be permitted to equate separately to 0 every term in which the modulus a does not satisfy the condition

$$a^n = a,$$

and to leave for the value of z the sum of the remaining terms.

As the nature of the demonstration of this Proposition is quite unaffected by the number of the terms in the second member, we will for simplicity confine ourselves to the supposition of there being four, and suppose that the moduli of the two first only, satisfy the index law.

We have then

$$w = a_1 t_1 + a_2 t_2 + a_3 t_3 + a_4 t_4, \quad (61),$$

with the relations $\quad a_1{}^n = a_1, \quad a_2{}^n = a_2,$

in addition to the two sets of relations connecting t_1, t_2, t_3, t_4, in accordance with (55) and (56).

Squaring (61), we have

$$w = a_1 t_1 + a_2 t_2 + a_3^2 t_3 + a_4^2 t_4,$$

and subtracting (61) from this,

$$(a_3{}^2 - a_3)\, t_3 + (a_4{}^2 - a_4)\, t_4 = 0\,;$$

and it being an hypothesis, that the coefficients of these terms do not vanish, we have, by Prop. 2,

$$t_3 = 0, \quad t_4 = 0, \quad (62),$$

whence (61) becomes

$$z = a_1 t_1 + a_2 t_2.$$

The utility of this Proposition will hereafter appear.

PROP. 4. The functions $t_1 t_2 . . t_r$ being mutually exclusive, we shall always have

$$\psi (a_1 t_1 + a_2 t_2 . . + a_r t_r) = \psi (a_1) t_1 + \psi (a_2) t_2 . . + \psi (a_r) t_r, \quad (63),$$

whatever may be the values of $a_1 a_2 . . a_r$ or the form of ψ.

Let the function $a_1 t_1 + a_2 t_2 . . + a_r t_r$ be represented by $\phi (xy . . .)$, then the moduli $a_1 a_2 . . a_r$ will be given by the expressions

$$\phi (11 . .), \quad \phi (10 . .), \quad (...) \quad \phi (00 . .).$$

Also $\psi (a_1 t_1 + a_2 t_2 . . + a_r t_r) = \psi \{\phi (xy . .)\}$

$$= \psi \{\phi (11 . .)\} \, xy . . + \psi \{\phi (10)\} \, x (1 - y) \cdots$$
$$+ \psi \{\phi (00)\} (1 - x)(1 - y) \cdots$$
$$= \psi (a_1) \, xy . . + \psi (a_2) \, x (1 - y) \cdots + \psi (a_r)(1 - x)(1 - y) \cdots$$
$$= \psi (a_1) \, t_1 + \psi (a_2) \, t_2 . . + \psi (a_r) \, t_r, \quad (64).$$

It would not be difficult to extend the list of interesting properties, of which the above are examples. But those which we have noticed are sufficient for our present requirements. The following Proposition may serve as an illustration of their utility.

PROP. 5. Whatever process of reasoning we apply to a single given Proposition, the result will either be the same Proposition or a limitation of it.

Let us represent the equation of the given Proposition under its most general form,

$$a_1 t_1 + a_2 t_2 . . + a_r t_r = 0, \quad (65),$$

resolvable into as many equations of the form $t = 0$ as there are moduli which do not vanish.

Now the most general transformation of this equation is

$$\psi (a_1 t_1 + a_2 t_2 . . + a_r t_r) = \psi (0), \quad (66),$$

provided that we attribute to ψ a perfectly arbitrary character, allowing it even to involve new elective symbols, having *any proposed relation* to the original ones.

F

The development of (66) gives, by the last Proposition,

$$\psi(a_1)\, t_1 + \psi(a_2)\, t_2 .\,. + \psi(a_r)\, t_r = \psi(0).$$

To reduce this to the general form of reference, it is only necessary to observe that since

$$t_1 + t_2 .\,. + t_r = 1,$$

we may write for $\psi(0)$,

$$\psi(0)\,(t_1 + t_2 .\,. + t_r),$$

whence, on substitution and transposition,

$$\{\psi(a_1) - \psi(0)\}\, t_1 + \{\psi(a_2) - \psi(0)\}\, t_2 .\,. + \{\psi(a_r) - \psi(0)\}\, t_r = 0.$$

From which it appears, that if a be any modulus of the
original equation, the corresponding modulus of the transformed
equation will be $\qquad \psi(a) - \psi(0).$

If $a = 0$, then $\psi(a) - \psi(0) = \psi(0) - \psi(0) = 0$, whence
there are no *new terms* in the transformed equation, and therefore there are no *new Propositions* given by equating its constituent members to 0.

Again, since $\psi(a) - \psi(0)$ may vanish without a vanishing,
terms may be wanting in the transformed equation which existed
in the primitive. Thus some of the constituent truths of the
original Proposition may entirely disappear from the interpre
tation of the final result.

Lastly, if $\psi(a) - \psi(0)$ do not vanish, it must either be
a numerical constant, or it must involve new elective symbols.
In the former case, the term in which it is found will give

$$t = 0,$$

which is one of the constituents of the original equation: in the
latter case we shall have

$$\{\psi(a - \psi(0)\}\, t = 0,$$

in which t has a limiting factor. The interpretation of this
equation, therefore, is a limitation of the interpretation of (65).

The purport of the last investigation will be more apparent to the mathematician than to the logician. As from any mathematical equation an infinite number of others may be deduced, it seemed to be necessary to shew that when the original equation expresses a logical Proposition, every member of the derived series, even when obtained by expansion under a functional sign, admits of exact and consistent interpretation.

OF THE SOLUTION OF ELECTIVE EQUATIONS.

In whatever way an elective symbol, considered as unknown, may be involved in a proposed equation, it is possible to assign its complete value in terms of the remaining elective symbols considered as known. It is to be observed of such equations, that from the very nature of elective symbols, they are necessarily linear, and that their solutions have a very close analogy with those of linear differential equations, arbitrary elective symbols in the one, occupying the place of arbitrary constants in the other. The method of solution we shall in the first place illustrate by particular examples, and, afterwards, apply to the investigation of general theorems.

Given $(1 - x) y = 0$, (All Ys are Xs), to determine y in terms of x.

As y is a function of x, we may assume $y = vx + v'(1 - x)$, (such being the expression of an arbitrary function of x), the moduli v and v' remaining to be determined. We have then

$$(1 - x) \left\{ vx + v'(1 - x) \right\} = 0,$$

or, on actual multiplication,

$$v'(1 - x) = 0 :$$

that this may be generally true, without imposing any restriction upon x, we must assume $v' = 0$, and there being no condition to limit v, we have

$$y = vx, \quad (67).$$

This is the complete solution of the equation. The condition that y is an elective symbol requires that v should be an elective symbol also (since it must satisfy the index law), its interpretation in other respects being arbitrary.

Similarly the solution of the equation, $xy = 0$, is

$$y = v(1 - x), \quad (68).$$

Given $(1 - x)zy = 0$, (All Ys which are Zs are Xs), to determine y.

As y is a function of x and z, we may assume

$$y = v(1 - x)(1 - z) + v'(1 - x)z + v''x(1 - z) + v'''zx.$$

And substituting, we get

$$v'(1 - x)z = 0,$$

whence $v' = 0$. The complete solution is therefore

$$y = v(1 - x)(1 - z) + v''x(1 - z) + v'''xz, \quad (69),$$

v', v'', v''', being arbitrary elective symbols, and the rigorous interpretation of this result is, that Every Y is *either* a not-X and not-Z, or an X and not-Z, or an X and Z.

It is deserving of note that the above equation may, in consequence of its linear form, be solved by adding the two particular solutions with reference to x and z; and replacing the arbitrary constants which each involves by an arbitrary function of the other symbol, the result is

$$y = x\phi(z) + (1 - z)\psi(x), \quad (70).$$

To shew that this solution is equivalent to the other, it is only necessary to substitute for the arbitrary functions $\phi(z)$, $\psi(x)$, their equivalents

$$wz + w'(1 - z) \text{ and } w''x + w'''(1 - x),$$

we get $y = wxz + (w' + w'')x(1 - z) + w'''(1 - x)(1 - z).$

In consequence of the perfectly arbitrary character of w' and w'', we may replace their sum by a single symbol w', whence

$$y = wxz + w'x(1 - z) + w'''(1 - x)(1 - z),$$

which agrees with (69).

The solution of the equation $wx(1 - y)z = 0$, expressed by arbitrary functions, is

$$z = (1 - w)\,\phi\,(xy) + (1 - x)\,\psi\,(wy) + y\chi\,(wx), \quad (71).$$

These instances may serve to shew the analogy which exists between the solutions of elective equations and those of the corresponding order of linear differential equations. Thus the expression of the ·integral of a partial differential equation, either by arbitrary functions or by a series with arbitrary coefficients, is in strict analogy with the case presented in the two last examples. To pursue this comparison further would minister to curiosity rather than to utility. We shall prefer to contemplate the problem of the solution of elective equations under its most general aspect, which is the object of the succeeding investigations.

To solve the general equation $\phi\,(xy) = 0$, with reference to y.

If we expand the given equation with reference to x and y, we have

$$\phi\,(00)\,(1 - x)\,(1 - y) + \phi\,(01)\,(1 - x)\,y + \phi\,(10)\,x\,(1 - y)$$
$$+\ \phi\,(11)\,xy = 0, \quad (72),$$

the coefficients $\phi\,(00)$ &c. being numerical constants.

Now the general expression of y, as a function of x, is

$$y = vx + v'\,(1 - x),$$

v and v' being unknown symbols to be determined. Substituting this value in (72), we obtain a result which may be written in the following form,

$$[\phi\,(10) + \{\phi\,(11) - \phi\,(10)\}\,v]\,x + [\phi\,(00) + \{\phi\,(00) - \phi\,(00)\}\,v']$$
$$(1 - x) = 0\,;$$

and in order that this equation may be satisfied without any way restricting the generality of x, we must have

$$\phi\,(10) + \{\phi\,(11) - \phi\,(10)\}\,v = 0,$$
$$\phi\,(00) + \{\phi\,(01) - \phi\,(00)\}\,v' = 0,$$

from which we deduce

$$v = \frac{\phi(10)}{\phi(10) - \phi(11)}, \qquad v' = \frac{\phi(00)}{\phi(01) - \phi(00)},$$

wherefore

$$y = \frac{\phi(10)}{\phi(10) - \phi(11)} \, x + \frac{\phi(00)}{\phi(00) - \phi(01)} \, (1 - x), \quad (73).$$

Had we expanded the original equation with respect to y only, we should have had

$$\phi(x\,0) + \{\phi(x\,1) - \phi(x\,0)\} \, y = 0 ;$$

but it might have startled those who are unaccustomed to the processes of Symbolical Algebra, had we from this equation deduced

$$y = \frac{\phi(x\,0)}{\phi(x\,0) - \phi(x\,1)},$$

because of the apparently meaningless character of the second member. Such a result would however have been perfectly lawful, and the expansion of the second member would have given us the solution above obtained. I shall in the following example employ this method, and shall only remark that those to whom it may appear doubtful, may verify its conclusions by the previous method.

To solve the general equation $\phi(xyz) = 0$, or in other words to determine the value of z as a function of x and y.

Expanding the given equation with reference to z, we have

$$\phi(xy\,0) + \{\phi(xy\,1) - \phi(xy\,0)\} \cdot z = 0 ;$$

$$\therefore z = \frac{\phi(xy\,0)}{\phi(xy\,0) - \phi(xy\,1)} \,\ldots (74),$$

and expanding the second member as a function of x and y by aid of the general theorem, we have

$$z = \frac{\phi(110)}{\phi(110) - \phi(111)} \, xy + \frac{\phi(100)}{\phi(100) - \phi(101)} \, x \, (1 - y)$$

$$+ \frac{\phi(010)}{\phi(010) - \phi(011)} \, (1 - x) \, y + \frac{\phi(000)}{\phi(000) - \phi(001)} \, (1 - x)(1 - y)$$

$$\ldots\ldots\ldots (75)$$

and this is the complete solution required. By the same method we may resolve an equation involving any proposed number of elective symbols.

In the interpretation of any general solution of this nature, the following cases may present themselves.

The values of the moduli $\phi(00)$, $\phi(01)$, &c. being constant, one or more of the coefficients of the solution may assume the form $\frac{0}{0}$ or $\frac{1}{0}$. In the former case, the indefinite symbol $\frac{0}{0}$ must be replaced by an arbitrary elective symbol v. In the latter case, the term, which is multiplied by a factor $\frac{1}{0}$ (or by any numerical constant except 1), must be separately equated to 0, and will indicate the existence of a subsidiary Proposition. This is evident from (62).

Ex. Given $x(1 - y) = 0$, All Xs are Ys, to determine y as a function of x.

Let $\phi(xy) = x(1 - y)$, then $\phi(10) = 1$, $\phi(11) = 0$, $\phi(01) = 0$, $\phi(00) = 0$; whence, by (73),

$$y = \frac{1}{1 - 0} x + \frac{0}{0 - 0}(1 - x)$$
$$= x + \frac{0}{0}(1 - x)$$
$$= x + v(1 - x), \quad (76),$$

v being an arbitrary elective symbol. The interpretation of this result is that the class Y consists of the entire class X with an indefinite remainder of not-Xs. This remainder is indefinite in the highest sense, *i.e.* it may vary from 0 up to the entire class of not-Xs.

Ex. Given $x(1 - z) + z = y$, (the class Y consists of the entire class Z, with such not-Zs as are Xs), to find Z.

Here $\phi(xyz) = x(1 - z) - y + z$, whence we have the following set of values for the moduli,

$\phi(110) = 0$, $\phi(111) = 0$, $\phi(100) = 1$, $\phi(101) = 1$,
$\phi(010) = -1$, $\phi(011) = 0$, $\phi(000) = 0$, $\phi(001) = 1$,

and substituting these in the general formula (75), we have

$$z = \frac{0}{0}xy + \frac{1}{0}x(1 - y) + (1 - x)y, \quad (77),$$

the infinite coefficient of the second term indicates the equation

$$x\,(1 - y) = 0, \text{ All Xs are Ys};$$

and the indeterminate coefficient of the first term being replaced by v, an arbitrary elective symbol, we have

$$z = (1 - x)\,y + vxy,$$

the interpretation of which is, that the class Z consists of all the Ys which are not Xs, and an *indefinite* remainder of Ys which are Xs. Of course this indefinite remainder may vanish. The two results we have obtained are logical inferences (not very obvious ones) from the original Propositions, and they give us all the information which it contains respecting the class Z, and its constituent elements.

Ex. Given $x = y\,(1 - z) + z\,(1 - y)$. The class X consists of all Ys which are not-Zs, and all Zs which are not-Ys: required the class Z.

We have

$$\phi\,(xyz) = x - y\,(1 - z) - z\,(1 - y),$$
$$\phi\,(110) = 0, \quad \phi\,(111) = 1, \quad \phi\,(100) = 1, \quad \phi\,(101) = 0,$$
$$\phi\,(010) = -1, \quad \phi\,(011) = 0, \quad \phi\,(000) = 0, \quad \phi\,(001) = -1;$$

whence, by substituting in (75),

$$z = x\,(1 - y) + y\,(1 - x), \quad (78),$$

the interpretation of which is, the class Z consists of all Xs which are not Ys, and of all Ys which are not Xs; an inference strictly logical.

Ex. Given $y\,\{1 - z\,(1 - x)\} = 0$, All Ys are Zs and not-Xs.

Proceeding as before to form the moduli, we have, on substitution in the general formulæ,

$$z = \tfrac{1}{0}\,xy + \tfrac{0}{0}\,x\,(1 - y) + y\,(1 - x) + \tfrac{0}{0}\,(1 - x)\,(1 - y),$$

or $z = y\,(1 - x) + vx\,(1 - y) + v'\,(1 - x)\,(1 - y)$

$$= y\,(1 - x) + (1 - y)\,\phi\,(x), \quad (79),$$

with the relation $\qquad xy = 0:$

from these it appears that No Ys are Xs, and that the class Z consists of all Ys which are not Xs, and of an indefinite remainder of not-Ys.

This method, in combination with Lagrange's method of indeterminate multipliers, may be very elegantly applied to the treatment of simultaneous equations. Our limits only permit us to offer a single example, but the subject is well deserving of further investigation.

Given the equations $x(1-z) = 0$, $z(1-y) = 0$, All Xs are Zs, All Zs are Ys, to determine the complete value of z with any subsidiary relations connecting x and y.

Adding the second equation multiplied by an indeterminate constant λ, to the first, we have

$$x(1-z) + \lambda z(1-y) = 0,$$

whence determining the moduli, and substituting in (75),

$$z = xy + \frac{1}{1-\lambda}\, x(1-y) + \tfrac{0}{0}(1-x)\, y, \quad (80),$$

from which we derive

$$z = xy + v(1-x)\, y,$$

with the subsidiary relation

$$x(1-y) = 0:$$

the former of these expresses that the class Z consists of all Xs that are Ys, with an indefinite remainder of not-Xs that are Ys; the latter, that All Xs are Ys, being in fact the conclusion of the syllogism of which the two given Propositions are the premises.

By assigning an appropriate meaning to our symbols, all the equations we have discussed would admit of interpretation in hypotheticals, but it may suffice to have considered them as examples of categoricals.

That peculiarity of elective symbols, in virtue of which every elective equation is reducible to a system of equations $t_1 = 0$, $t_2 = 0$, &c., so constituted, that all the binary products $t_1 t_2$, $t_1 t_3$, &c., vanish, represents a general doctrine in Logic with reference to the ultimate analysis of Propositions, of which it may be desirable to offer some illustration.

Any of these constituents t_1, t_2, &c. consists only of factors of the forms x, y,...$1 - w$, $1 - z$, &c.   In categoricals it therefore represents a compound class, *i.e.* a class defined by the presence of certain qualities, and by the absence of certain other qualities.

Each constituent equation $t_1 = 0$, &c. expresses a denial of the existence of some class so defined, and the different classes are mutually exclusive.

Thus all categorical Propositions are resolvable into a denial of the existence of certain compound classes, no member of one such class being a member of another.

The Proposition, All Xs are Ys, expressed by the equation $x (1 - y) = 0$, is resolved into a denial of the existence of a class whose members are Xs and not-Ys.

The Proposition Some Xs are Ys, expressed by $v = xy$, is resolvable as follows.   On expansion,

$$v - xy = vx (1 - y) + vy (1 - x) + v (1 - x) (1 - y) - xy (1 - v);$$
$$\therefore vx (1 - y) = 0, \; vy (1 - x) = 0, \; v (1 - x) (1 - y) = 0, \; (1 - v) xy = 0.$$

The three first imply that there is no class whose members belong to a certain unknown Some, and are 1st, Xs and not Ys ; 2nd, Ys and not Xs ; 3rd, not-Xs and not-Ys.   The fourth implies that there is no class whose members are Xs and Ys without belonging to this unknown Some.

From the same analysis it appears that *all hypothetical Propositions may be resolved into denials of the coexistence of the truth or falsity of certain assertions.*

Thus the Proposition, If X is true, Y is true, is resolvable· by its equation $x (1 - y) = 0$, into a denial that the truth of X and the falsity of Y coexist.

And the Proposition Either X is true, or Y is true, members exclusive, is resolvable into a denial, first, that X and Y are both true ; secondly, that X and Y are both false.

But it may be asked, is not something more than a system of negations necessary to the constitution of an affirmative Proposition ? is not a positive element required ?   Undoubtedly there is need of one; and this positive element is supplied in categoricals by the assumption (which may be regarded as a prerequisite of reasoning in such cases) that there *is* a Universe of conceptions, and that each individual it contains either belongs to a proposed class or does not belong to it; in hypotheticals, by the assumption (equally prerequisite) that there is a Universe of conceivable cases, and that any given Proposition is either true or false. Indeed the question of the existence of conceptions (εἰ ἔστι) is preliminary to any statement of their qualities or relations (τί ἔστι).—*Aristotle, Anal. Post.* lib. II. cap. 2.

It would appear from the above, that Propositions may be regarded as resting at once upon a positive and upon a negative foundation. Nor is such a view either foreign to the spirit of Deductive Reasoning or inappropriate to its Method; the latter ever proceeding by limitations, while the former contemplates the particular as derived from the general.

Demonstration of the Method of Indeterminate Multipliers, as applied to Simultaneous Elective Equations.

To avoid needless complexity, it will be sufficient to consider the case of three equations involving three elective symbols, those equations being the most general of the kind. It will be seen that the case is marked by every feature affecting the character of the demonstration, which would present itself in the discussion of the more general problem in which the number of equations and the number of variables are both unlimited.

Let the given equations be

$$\phi\,(xyz) = 0, \quad \psi\,(xyz) = 0, \quad \chi\,(xyz) = 0, \quad (1).$$

Multiplying the second and third of these by the arbitrary constants h and k, and adding to the first, we have

$$\phi\,(xyz) + h\,\psi\,(xyz) + k\,\chi\,(xyz) = 0, \quad (2);$$

and we are to shew, that in solving this equation with reference to any variable z by the general theorem (75), we shall obtain not only the general value of z independent of h and k, but also any subsidiary relations which may exist between x and y independently of z.

If we represent the general equation (2) under the form $F(xyz) = 0$, its solution may by (75) be written in the form

$$z = \cfrac{xy}{1 - \cfrac{F(111)}{F(110)}} + \cfrac{x(1-y)}{1 - \cfrac{F(101)}{F(100)}} + \cfrac{y(1-x)}{1 - \cfrac{F(011)}{F(010)}} + \cfrac{(1-x)(1-y)}{1 - \cfrac{F(001)}{F(000)}};$$

and we have seen, that any one of these four terms is to be equated to 0, whose modulus, which we may represent by M, does not satisfy the condition $M^n = M$, or, which is here the same thing, whose modulus has any other value than 0 or 1.

Consider the modulus (suppose M_1) of the first term, viz. $\cfrac{1}{1 - \cfrac{F(111)}{F(110)}}$, and giving to the symbol F its full meaning,

we have
$$M_1 = \cfrac{1}{1 - \cfrac{\phi(111) + h\psi(111) + k\chi(111)}{\phi(110) + h\psi(110) + k\chi(110)}}.$$

It is evident that the condition $M_1^n = M_1$ cannot be satisfied unless the right-hand member be independent of h and k; and in order that this may be the case, we must have the function $\dfrac{\phi(111) + h\psi(111) + k\chi(111)}{\phi(110) + h\psi(110) + k\chi(110)}$ independent of h and k.

Assume then
$$\frac{\phi(111) + h\psi(111) + k\chi(111)}{\phi(110) + h\psi(110) + k\chi(110)} = c,$$

c being independent of h and k; we have, on clearing of fractions and equating coefficients,

$$\phi(111) = c\phi(110), \quad \psi(111) = c\psi(110), \quad \chi(111) = c\chi(110);$$

whence, eliminating c,

$$\frac{\phi(111)}{\phi(110)} = \frac{\psi(111)}{\psi(110)} = \frac{\chi(111)}{\chi(110)},$$

being equivalent to the triple system

$$\left.\begin{aligned}\phi(111)\,\psi(110) - \phi(110)\,\psi(111) = 0\\ \psi(111)\,\chi(110) - \psi(110)\,\chi(111) = 0\\ \chi(111)\,\phi(110) - \chi(110)\,\psi(111) = 0\end{aligned}\right\} \quad (3);$$

and it appears that if any one of these equations is not satisfied, the modulus M_1 will not satisfy the condition $M_1{}^n = M_1$, whence the first term of the value of z must be equated to 0, and we shall have

$$xy = 0,$$

a relation between x and y independent of z.

Now if we expand in terms of z each pair of the primitive equations (1), we shall have

$$\phi(xy0) + \{\phi(xy1) - \phi(xy0)\}\,z = 0,$$
$$\psi(xy0) + \{\psi(xy1) - \psi(xy0)\}\,z = 0,$$
$$\chi(xy0) + \{\chi(xy1) - \chi(xy0)\}\,z = 0,$$

and successively eliminating z between each pair of these equations, we have

$$\phi(xy1)\,\psi(xy0) - \phi(xy0)\,\psi(xy1) = 0,$$
$$\psi(xy1)\,\chi(xy0) - \psi(xy0)\,\chi(xy1) = 0,$$
$$\chi(xy1)\,\phi(xy0) - \chi(xy0)\,\phi(xy1) = 0,$$

which express all the relations between x and y that are formed by the elimination of z. Expanding these, and writing in full the first term, we have

$$\{\phi(111)\,\psi(110) - \phi(110)\,\psi(111)\}\,xy + \&\text{c.} = 0,$$
$$\{\psi(111)\,\chi(110) - \psi(110)\,\chi(111)\}\,xy + \&\text{c.} = 0,$$
$$\{\chi(111)\,\phi(110) - \chi(110)\,\phi(111)\}\,xy + \&\text{c.} = 0:$$

and it appears from Prop. 3, that if the coefficient of xy in any of these equations does not vanish, we shall have the equation

$$xy = 0;$$

but the coefficients in question are the same as the first members of the system (3), and the two sets of conditions exactly agree. Thus, as respects the first term of the expansion, the method of indeterminate coefficients leads to the same result as ordinary elimination; and it is obvious that from their similarity of form, the same reasoning will apply to all the other terms.

Suppose, in the second place, that the conditions (3) are satisfied so that M_1 is independent of h and k. It will then indifferently assume the equivalent forms

$$M_1 = \cfrac{1}{1 - \cfrac{\phi(111)}{\phi(110)}} = \cfrac{1}{1 - \cfrac{\psi(111)}{\psi(110)}} = \cfrac{1}{1 - \cfrac{\chi(111)}{\chi(110)}}.$$

These are the exact forms of the first modulus in the expanded values of z, deduced from the solution of the three primitive equations singly. If this common value of M_1 is 1 or $\frac{0}{0} = v$, the term will be retained in z; if any other constant value (except 0), we have a relation $xy = 0$, not given by elimination, but deducible from the primitive equations singly, and similarly for all the other terms. Thus in every case the expression of the subsidiary relations is a necessary accompaniment of the process of solution.

It is evident, upon consideration, that a similar proof will apply to the discussion of a system indefinite as to the number both of its symbols and of its equations.

POSTSCRIPT.

SOME additional explanations and references which have occurred to me during the printing of this work are subjoined.

The remarks on the connexion between Logic and Language, p. 5, are scarcely sufficiently explicit. Both the one and the other I hold to depend very materially upon our ability to form general notions by the faculty of abstraction. Language is an instrument of Logic, but not an indispensable instrument.

To the remarks on Cause, p. 12, I desire to add the following : Considering Cause as an invariable antecedent in Nature, (which is Brown's view), whether associated or not with the idea of Power, as suggested by Sir John Herschel, the knowledge of its existence is a knowledge which is properly expressed by the word *that* (τὸ ὅτι), not by *why* (τὸ διότι). It is very remarkable that the two greatest authorities in Logic, modern and ancient, agreeing in the latter interpretation, differ most widely in its application to Mathematics. Sir W. Hamilton says that Mathematics exhibit only the *that* (τὸ ὅτι): Aristotle says, The *why* belongs to mathematicians, for they have the demonstrations of Causes. *Anal. Post.* lib. I., cap. XIV. It must be added that Aristotle's view is consistent with the sense (albeit an erroneous one) which in various parts of his writings he virtually assigns to the word Cause, viz. an antecedent in Logic, a sense according to which the premises might be said to be the cause of the conclusion. This view appears to me to give even to his physical inquiries much of their peculiar character.

Upon reconsideration, I think that the view on p. 41, as to the presence or absence of a medium of comparison, would readily follow from Professor De Morgan's doctrine, and I therefore relinquish all claim to a discovery. The mode in which it appears in this treatise is, however, remarkable.

I have seen reason to change the opinion expressed in pp. 42, 43. The system of equations there given for the expression of Propositions in Syllogism is *always* preferable to the one before employed—first, in generality—secondly, in facility of interpretation.

In virtue of the principle, that a Proposition is either true or false, every elective symbol employed in the expression of hypotheticals admits only of the values 0 and 1, which are the only quantitative forms of an elective symbol. It is in fact possible, setting out from the theory of Probabilities (which is purely quantitative), to arrive at a system of methods and processes for the treatment of hypotheticals exactly similar to those which have been given. The two systems of elective symbols and of quantity osculate, if I may use the expression, in the points 0 and 1. It seems to me to be implied by this, that unconditional truth (categoricals) and probable truth meet together in the constitution of contingent truth, (hypotheticals). The general doctrine of elective symbols and all the more characteristic applications are quite independent of any quantitative origin.

THE END.

ERRATA.

Page 5, note, *for* vi. *read* iv.
" 8, line 31, *for* first *read* fact.
" 17, " 12, *for* abstraction *read* election
" 53, " 12, *for* numbers *read* members.
" 66, " 12 and 32, *for* z *read* w.
" 80, " 27, *for* 3 *read* 2.

Printed in the United States
By Bookmasters